T0182026

Studies in Computational Intelligence

Volume 492

Series Editor

J. Kacprzyk, Warsaw, Poland

For further volumes:
http://www.springer.com/series/7092

Roger Lee
Editor

Software Engineering, Artificial Intelligence, Networking and Parallel/Distributed Computing

Editor
Roger Lee
Software Engineering and Information Technology Institute
Central Michigan University
Michigan
USA

ISSN 1860-949X ISSN 1860-9503 (electronic)
ISBN 978-3-319-03272-6 ISBN 978-3-319-00738-0 (eBook)
DOI 10.1007/978-3-319-00738-0
Springer Cham Heidelberg New York Dordrecht London

Printed on acid-free paper

Springer is part of Springer Science+Business Media (www.springer.com)

Preface

The purpose of the 14th ACIS/IEEE International Conference on Software Engineering, Artificial Intelligence, Networking and Parallel/Distributed Computing (SNPD 2013), held in Honolulu, Hawaii, USA on July 1–3, 2013 is aimed at bringing together researchers and scientists, businessmen and entrepreneurs, teachers and students to discuss the numerous fields of computer science, and to share ideas and information in a meaningful way. Our conference officers selected the best 17 papers accepted for presentation at the conference in order to publish them in this volume. The papers were chosen based on review scores submitted by members of the program committee and underwent further rounds of rigorous review.

In chapter 1, Nakashima et al. propose a sensor, which detects person localization without privacy offending, applying brightness distribution. In the proposed design, the sensor is constructed with a line sensor and cylindrical lens to obtain one-dimensional brightness distribution. Comparing with conventional line sensors, CMOS area sensors are with low cost, and high sensitivity. Therefore, in the proposed sensor, the CMOS area sensor is applied as covered in certain areas physically, so that it behaves as a line sensor. The proposed sensor is able to obtain one-dimensional horizontal brightness distribution that is approximately equal to integration value of each vertical pixel line of two-dimensional image.

In chapter 2, Irosh Fernando, Frans Henskens and Martin Cohen propose a model attempting to overcome the main limitations of approximating reasoning in medical diagnosis. Unfortunately, most current approaches have not been able to model medical diagnostic reasoning sufficiently, or in a clinically intuitive way.

In chapter 3, Shigeru Nakayama and Peng Gang modify the adiabatic quantum computation and propose to solve Deutsch-Jozsa problem more efficiently by a method with higher observation probability. Adiabatic quantum computation has been proposed as a quantum algorithm with adiabatic evolution to solve combinatorial
optimization problem, then it has been applied to many problems like satisfiability problem.

In chapter 4, Tatsuya Sakato, Motoyuki Ozeki, and Natsuki Oka propose a computational model of imitation and autonomous behavior. In the model, an agent can reduce its learning time through imitation. They extend the model to continuous spaces, and add a function for selecting a target action for imitation from observed actions to the model. By these extension and adaptation, the model comes to adapt to more complex environment.

In chapter 5, Jianfei Zhang et al. propose analyze the main factors that in?uence the performance of CTC(N) and present an improved contentiontolerant switch architecture - Diagonalized Contention-Tolerant Crossbar Switch, denoted as DiaCTC(N). They recently proposed an innovative agile crossbar switch architecture called
Contention-Tolerant Crossbar Switch, denoted as CTC(N), where N is the number of input/output ports. CTC(N) can tolerate output contentions instead of resolving them by complex hardware, which makes CTC(N) simpler and more scalable than conventional crossbar switches.

In chapter 6, Hajime Murai introduces four methods of quantitative analysis for the interpretation of the Bible in a scientific manner. The methods are citation analysis for interpreters' texts, vocabulary analysis for translations, variant text analysis for canonical texts, and an evaluation method for rhetorical structure. Furthermore, these algorithms are implemented for Java-based software.

In chapter 7, Djelloul Bouchiha et al. present an approach for annotating Web services. The approach consists of two processes, namely the categorization and matching. Both processes use ontology matching techniques. In particular, the two processes use similarity measures between entities, strategies for computing similarities between sets and a threshold corresponding to the accuracy. Thus, an internal comparative study has been done to answer the questions: which strategy is appropriate to this approach? Which measure gives best results? And which threshold is optimum for the selected measure and strategy? An external comparative study is also useful to prove the efficacy of this approach compared to existing annotation approaches.

In chapter 8, Xulong Tang, Hong An, Gongjin Sun, and Dongrui Fan introduce VCBench, a Video Coding Benchmark suite which is built up from a wide range of video coding applications. Firstly, typical codecs are selected (X264, XVID, VP8) according to the popularity, coding efficiency, compression quality, and source accessibility. Secondly, hotspots are extracted from coding process as kernels. VCBench mainly focuses on Transformation, Quantization and Loop filter. All of the extracted kernels are single-threaded version without any architecture-specific optimizations, such as SIMD. Besides, inputs in three different sizes are provided to control running time. Finally, to better understand intrinsic characteristics of video coding application, we analyze both computational and memory characteristics, and further provide insights into architectural design which can improve the performance of this kind of applications.

In chapter 9, Elena Acevedo, Antonio Acevedo, Fabiola Martínez, and Ángel Martínez describe a novel method for encrypting monochromatic images using an associative approach. The image is divided in blocks which are used to build max and min Alpha-Beta associative memories. The key is private and it depends on the number of blocks. The main advantage of this method is that the encrypted image does not have the same size than the original image; therefore, since the beginning the adversary cannot know what the image means.

In chapter 10, Jinhee Park, Hyeon-Jeong Kim, and Jongmoon Baik deal with reliability evaluation issues in the web environment and compare with each other in terms of failure data collection methods, reliability evaluation techniques, and validation schemes. We also evaluate them based on hypothetical execution scenarios, analyze the strengths or weaknesses of each technique, and identify the remaining open problems.

In chapter 11, Jie Lin et al. Lu develop the distributed energy routing protocols for the smart grid. In particular, they first develop the Global Optimal Energy Routing Protocol (GOER), which e?ciently distributes energy with minimum transmission overhead. Considering that the computation overhead of GOER limits its use in large-scale grids, they then develop the Local Optimal Energy Routing Protocol (LOER) for large-scale grids. The basic idea of LOER is to divide the grid into multiple regions and adopt a multiple layer optimal strategy to reduce the energy distribution overhead while preserving the low computation overhead. Through extensive theoretical analysis and simulation experiments, their data shows that their developed protocols can provide higher energy distribution e?ciency in comparison with the other protocols.

In chapter 12, Yoondeok Ju and Moonju Park present a framework for efficient use of multicore CPU in embedded systems. The proposed framework monitors the usage of the computing resources such as CPU cores, memory, network, and the number of threads. Then it manages the number of CPU cores to be assigned to the application using the resource usage hints. They have tested the framework using SunSpider benchmark with FireFox and Midori Web browsers on an embedded system with Exynos4412 quad-core. Experimental results show that by managing the core assignment and frequency scaling, we can improve the energy efficiency along with the performance.

In chapter 13, Hideo Hirose, Junaida Binti Sulaiman, Masakazu Tokunaga apply the matrix decomposition method to predict the amount of seasonal rainfalls. Applying the method to the case of Indian rainfall data from 1871 to 2011, they have found that the early detection and prediction for the extreme-value of the monthly rainfall can be attained. Using the newly introduced accuracy evaluation criterion, risky, they can see that the matrix decomposition method using cylinder-type matrix provides the comparative accuracy to the artificial neural network result which has been conventionally used.

In chapter 14, Shunsuke Akai, Teruhisa Hochin and Hiroki Nomiya propose an impression evaluation method considering the vagueness of Kansei. This evaluation method makes a subject evaluate impression spatially. A method for analyzing the evaluation results has been proposed and the results of analysis have been shown. This analysis method shows average values and coefficients of variation of scores of the evaluation results spatially.

In chapter 15, Naotaka Oda et al. challenges a new approach to contribute to the student counseling field by giving a representation of counseling records. By giving a representation, effective analysis and processing of cases are possible. First, they give a formal representation of counseling cases based on observation that persons and their network and changes of the network are important structure of cases. Second, they try to capture characteristics of cases by giving attributes and transitions of relations that compose networks. Third, a similarity measure is defined for cases based on the formal representation and attributes. These proposal are examined by counseling cases, which are prepared by rearranging real cases.

In chapter 16, Khan Md Anwarus Salam, Hiroshi Uchida, Setsuo Yamada, and Nishio Tetsuro develop the web based UNL graph editor which visualizes UNL expression and the users can edit the graphs interactively. The Universal Networking Language (UNL) is an artificial language which enables computers to process knowledge and information across language barriers and cultural differences. Universal Words (UWs) constitute the vocabulary of UNL. However, from the UNL expression it is difficult for humans to visualize the UNL graph to edit it interactively.

In chapter 17, Chi-Min Oh and Chil-Woo Lee propose a method which can improve MRF-Particle filters used to solve the hijacking problem; independent particle filters for tracking each object can be kidnapped by a neighboring target which has higher likelihood than that of real target. In the method the motion model built by Markov random field (MRF) has been usefully applied for avoiding hijacking by lowering the weight of particles which are very close to any neighboring target. The MRF unary and pairwise potential functions of neighboring targets are defined as the penalty function to lower the particle's weights. And potential function can be reused for defining action likelihood which can measure the motion of object group.

It is our sincere hope that this volume provides stimulation and inspiration, and that it will be used as a foundation for works to come.

July 2013

Satoshi Takahashi
Program Chair

Contents

List of Contributors

Antonio Acevedo
Instituto Politécnico Nacional, Mexico
macevedo@ipn.mx

Elena Acevedo
Instituto Politécnico Nacional, Mexico
eacevedo@ipn.mx

Shunsuke Akai
Kyoto Institute of Technology, Japan
m2622001@edu.kit.ac.jp

Abdullah Alghamdi
KSU, Saudi Arabia
Ghamdi@ksu.edu.sa

Khalid Alnafjan
KSU, Saudi Arabia
Alnafjan@ksu.edu.sa

Hong An
USTC, China
han@ustc.edu.cn

Jongmoon Baik
Korea Advanced Institute of Science and Technology, Republic of Korea
jbaik@kaist.ac.kr

Djelloul Bouchiha
Djillali Liabes University of Sidi Bel Abbes, Algeria
bouchiha.dj@gmail.com

Martin Cohen
The Mater Hospital, Hunter New England Area Health Service, Australia
martin.cohen@hnehealth.nsw.gov.au

Djihad Djaa
University Dr. Taher Moulay, Algeria
djihad22@hotmail.fr

Dongrui Fan
Chinese Academy of Sciences, China
fandr@ict.ac.cn

Zhiyi Fang
Jilin University, China

Irosh Fernando
University of Newcastle, Australia
irosh.fernando@uon.edu.au

Peng Gang
Huizhou University, China
peng@hzu.edu.cn

David Griffith
National Institute of Standards and Technology (NIST), USA
david.griffith@nist.gov

Frans Henskens
University of Newcastle, Australia
Frans.henskens@newcastle.edu.au

Hideo Hirose
Kyushu Institute of Technology, Japan
hirose@ces.kyutech.ac.jp

Teruhisa Hochin
Kyoto Institute of Technology, Japan
hochin@kit.ac.jp

Nobuhiro Inuzuka
Nagoya Institute of Technology, Japan
inuzuka@nitech.ac.jp

Yoondeok Ju
Incheon National University, Korea
dbsejr21@gmail.com

Hyeon-Jeong Kim
Korea Advanced Institute of Science and Technology, Republic of Korea
hjkim@se.kaist.ac.kr

Yuhki Kitazono
Kitakyushu National College of Technology, Japan

Chil-Woo Lee
Chonnam National University, Korea
leecw@chonnam.ac.kr

Jie Lin
Xi'an Jiaotong University, China
Dr.linjie@stu.xjtu.edu.cn

Chao Lu
Towson University, USA
clu@towson.edu

Mimoun Malki
Djillali Liabes University of Sidi Bel Abbes, Algeria
malki@univ-sba.dz

Ángel Martínez
Instituto Politécnico Nacional, Mexico
josekun13@gmail.com

Fabiola Martínez
Instituto Politécnico Nacional, Mexico
fmartinezzu@ipn.mx

Shenglin Mu
Hiroshima National College of Maritime Technology, Japan

Hajime Murai
Tokyo Institute of Technology, Japan
h_murai@valdes.titech.ac.jp

Atsuko Mutoh
Nagoya Institute of Technology, Japan
mutoh@nitech.ac.jp

Shorta Nakashima
Yamaguchi University, Japan
s-naka@yamaguchiu.ac.jp

Shigeru Nakayama
Kagoshima University, Japan
shignaka@ibe.kagoshima-u.ac.jp

Aya Nishimura
Nagoya Institute of Technology, Japan
nishimura@nous.nitech.ac.jp

Hiroki Nomiya
Kyoto Institute of Technology, Japan
nomiya@kit.ac.jp

Naotaka Oda
Nagoya Institute of Technology, Japan
oda@nous.nitech.ac.jp

Chi-Min Oh
Chonnam National University, Korea
sapeyes@image.chonnam.ac.kr

Natsuki Oka
Kyoto Institute of Technology, Japan
natg@kit.ac.jp

Motoyuki Ozeki
Kyoto Institute of Technology, Japan
ozeki@kit.ac.jp

Jinhee Park
Korea Advanced Institute of Science and Technology, Republic of Korea
jh_park@kaist.ac.kr

Moonju Park
Incheon National University, Korea
mpark@incheon.ac.kr

Guannan Qu
Jilin University, China

Tatsuya Sakato
Kyoto Institute of Technology, Japan
sakato@ii.is.kit.ac.jp

Khan Md Anwarus Salam
The University of Electro-Communications, Japan
kmanwar@gmail.com

Takuya Seko
Nagoya Institute of Technology, Japan
seko@nous.nitech.ac.jp

Seiichi Serikawa
Kyushu Institute of Technology, Japan

Okabe Shintaro
Yamaguchi University, Japan

Junaida Binti Sulaiman
Kyushu Institute of Technology, Japan

Gongjin Sun
USTC, China
hilcutz@mail.ustc.edu.cn

Kanya Tanaka
Yamaguchi University, Japan

Xulong Tang
USTC, China
tangxl@mail.ustc.edu.cn

Nishio Tetsuro
The University of Electro-Communications, Japan
nishino@uec.ac.jp

Masakazu Tokunaga
Kyushu Institute of Technology, Japan

Hiroshi Uchida
UNDL Foundation, Japan
uchida@undl.org

Yuji Wakasa
Yamaguchi University, Japan

Guobin Xu
Towson University, USA
tigerguobin@gmail.com

Setsuo Yamada
NTT Corporation, Japan
yamada.setsuo@lab.ntt.co.jp

Xinyu Yang
Xi'an Jiaotong University, China
yxyphd@mail.xjtu.edu.cn

Wei Yu
Towson University, USA
wyu@towson.edu

Jianfei Zhang
Jilin University, China
qu.guannan@hotmail.com

Xiaohui Zhao
Jilin University, China

S.Q. Zheng
University of Texas at Dallas, USA

Restroom Human Detection Using One-Dimensional Brightness Distribution Sensor

Shorta Nakashima, Shenglin Mu, Okabe Shintaro, Kanya Tanaka, Yuji Wakasa, Yuhki Kitazono, and Seiichi Serikawa

Abstract. As aging society problem goes serious; systems to confirm safety of elders in daily life are expected. In this paper, a sensor, which detects person localization without privacy offending, applying brightness distribution is realized. In the proposed design, the sensor is constructed with a line sensor and cylindrical lens to obtain one-dimensional brightness distribution. Comparing with conventional line sensors, CMOS area sensors are with low cost, and high sensitivity. Therefore, in the proposed sensor, the CMOS area sensor is applied as covered in certain areas physically, so that it behaves as a line sensor. The proposed sensor is able to obtain one-dimensional horizontal brightness distribution that is approximately equal to integration value of each vertical pixel line of two-dimensional image. By employing this method, the information of the target person's position and movement status can be obtained without using two-dimensional detail texture image.

1 Introduction

Aging society is a serious problem which the whole Japan has to face to. The increasing number of elders is putting more and more burden on the workers in medical and welfare fields. Especially, in daily life of elders, much more workers are

Shorta Nakashima · Okabe Shintaro · Kanya Tanaka · Yuji Wakasa
Yamaguchi University, 2-16-1 Tokiwadai, Ube, Yamaguchi, Japan
e-mail: `s-naka@yamaguchi-u.ac.jp`

Shenglin Mu
Hiroshima National College of Maritime Technology, 4272-1 Higashino, Osakikamishima-cho, Toyota-kun, Hiroshima, Japan

Yuhki Kitazono
Kitakyushu National College of Technology, 5-20-1 Shii, Kokuraminami, Kitakyushu, Fukuoka, Japan

Seiichi Serikawa
Kyushu Institute of Technology, 1-1 Sensui-cho, Tobata, Kitakyushu, Fukuoka, Japan

R. Lee (Ed.): *SNPD*, SCI 492, pp. 1–10.
DOI: 10.1007/978-3-319-00738-0_1

required to take care of them, even just to confirm their safety. Usually, in daily care of elders, falling down in restroom causes serious accidents. Surveillance systems are necessary and important for relieving the burden of care workers in preventing elders from accidents stated above. However, there are several limitations of conventional surveillance systems with video cameras, such as high cost and hard to set. Especially, because of privacy protection policy they are not able to apply in restroom, where falling down accidents usually happen. Although, for solving that problem, there are some systems with video image deteriorated designed have been proposed, the systems with cameras are still not easy to be employed [1, 2, 3]. Therefore, the system, which can detect a person's position and movement status without video or image information, is expected in medical and welfare field.

In previous research, pyroelectric type temperature sensors are generally used in detecting a person's presence by infrared rays. Limitation of the sensor is that it cannot distinguish the person's position and/or movement status, but only presence [4]. In order to solve the problem, a sensor which is able to get localization of target person without privacy offending is proposed in our previous work introduced in [5]. It is considered that target person can be localized by using the integration value of each vertical pixel line of two-dimensional image instead of one-dimensional brightness distribution obtained by combining the line sensor and the cylindrical lens, because these two are thought to be the same approximately in theory. Simulation results in previous research showed that the proposed method is effective at confirming a person's presence, position, and movement status. Especially, it is attractive for welfare application that the method is able to distinguish the target person's standing upright or falling down status based on one-dimensional brightness distribution with no privacy offending at all.

For evaluating the proposed method in not only simulation but also field test, the sensor system, which is proposed according to the method in previous research for person's localization without privacy offending, is realized and varified in this paper. In this research, the sensor was constructed and the field tests for the sensor were implemented. According to the experimental results, it can be stated that target person's position could be detected from only one-dimensional brightness distribution without considering two-dimensional information fetched by the proposed sensor sytem. The proposed system is effective in real application in restroom. The paper is organized as follows. After this introduction section, Sect. 2 gives the introduction about the privacy-preventive sensor. In Sect. 3, the basic structure of the Obrid-Sensor proposed in this research is introduced. The acquisition model of brightness distribution is stated in Sect. 4. After that, the experiments are studied in Sect. 5. The conclusions are given in the last section.

2 Privacy-Preventive Sensor

The sensor which is able to detect a person's position and state without privacy offending was introduced in our previous work, named as Obrid-Sensor [5]. The Obrid-Sensor can obtain one-dimensional brightness distribution by combining signal fetched by line sensor and rod lens (the detail about the design is given in Sect. 4). The state of the target person can be estimated using the one-dimensional brightness distribution by the LVQ algorithm [6]. In this method, in order to obtain the optimum parameters in LVQ, the S-System [7, 8], which is one of the extended models of GA [9] and GP [10], is employed. By employing the method, a privacy-preserving sensor, which can distinguish target person's states between standing and falling down without using cameras or other video equipments, can be constructed. In addition, the application of the S-System provides good parameters despite a lack of experience and familiarity, and a discrimination circuit was created using LVQ.

Based on the simulation results using the brightness distribution from a camera captured image, the effectiveness of our design was verified. In this paper, the designed sensor system was realized according to the proposed method. The proposed scheme was confirmed by experimental results in field tests.

3 Basic Structure of Obrid-Sensor

As we see in Fig. 1 it is the basic structure of Obrid-Sensor. The sensor is constructed with a CMOS sensor and a rod lens as we see in the figure. Comparing with expensive line sensors, CMOS area sensors are with features of low cost and high sensitivity. The application of them is increasing according to recent rapid advancement in CMOS technology. Therefore, in this research, a CMOS image sensor is designed to be covered by black covers that have a line shaped window over the center row, so that it will behave as a line sensor in practice. The rod lens has the same properties as the cylindrical lens which is introduced in [5].

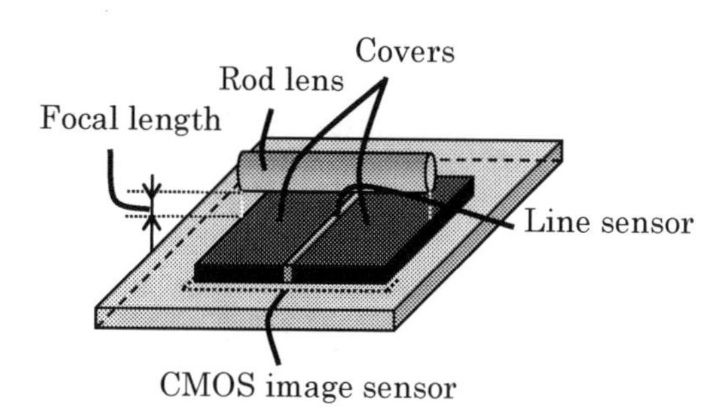

Fig. 1 Structure of Obrid-Sensor

4 Acquisition Model of Brightness Distribution

The proposed sensor is designed to detect a target person's state without applying video or image information. The theoretical structure of Obrid-Sensor is shown in Fig. 2. Figure 2(a), (b), and (c) show its overall view, side view and top view, respectively. As we see in Fig. 2(b), the rod lens is with the shape of a flat panel, and the light radiation from a physical object is input into the sensor even if a vertical

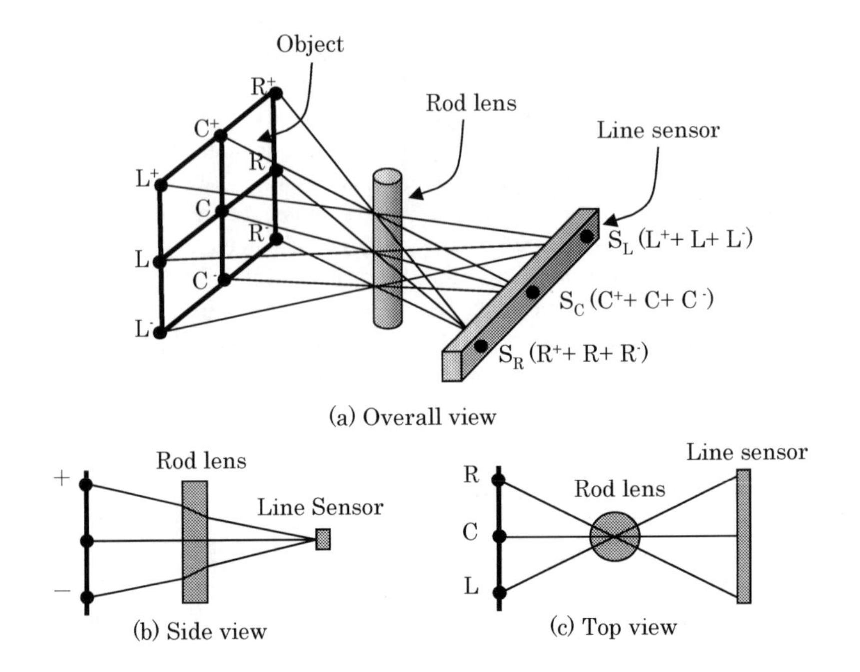

Fig. 2 Setting in Obrid-Sensor

position is different. In other words, the distribution of vertical light is integrated. In addition, as shown in Fig. 2(c), the rod lens is a convex lens when it is viewed from the top. Therefore, when the physical object's horizontal position is different, the position projected on the sensor will also be different. Summarizing information stated above, Fig. 2(a) shows the light radiated from the physical object's R^+, R, R^- enters the SR point of the line sensor that radiated from C^+, C, C^-, enters the SC point, and L^+, L, L^- enters the SL point. In other words, the output from the line sensor indicates how the vertical integration of the light from the object distributes in the horizontal direction. Thus, such privacy-preserving sensor can be achieved without using cameras or using a large scale device.

5 Experiments and Study

In order to verify the usefulness of the proposed system, a group of experiments were implemented and studied. Firstly, target person's localization experiments were implemented for verification of the sensor's accuracy. In the experiments, the brightness distribution of the sensor was measured when the target person moved in a fixed range. The experimental procedure can be stated as follows. First, the brightness distribution of the background where the person was not present was acquired. Next, the brightness distribution of the area where the person was present was ac-

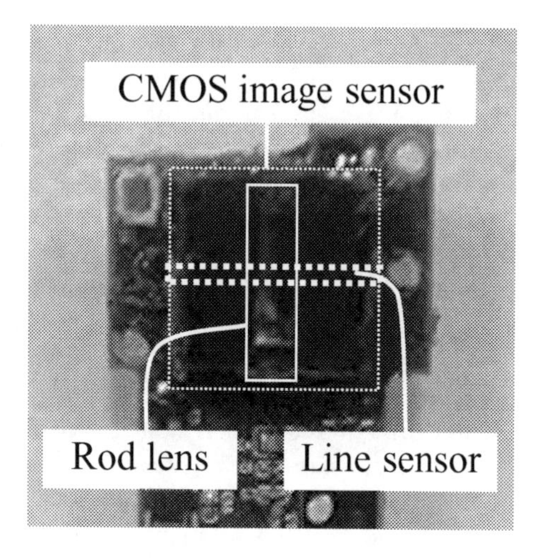

Fig. 3 Appearance of Obrid-Sensor

quired. Then the difference between the background and the brightness distribution of the object can be calculated. Because the difference of brightness distribution changed according to target person's movement, it is thought that the person's position information was included in the difference value.

In this research, the proposed sensor for field tests was realized by a rod lens (SAKAI-G rod lens φ 3 $\times$ 110 mm) and a CMOS image sensor (Logicool Qcam Pro 9000) as shown in Fig. 3. In the experiment, according to the basic design shown in Fig. 4, the usable range r for person localization was measured when the condition of the distance d between object and Obrid-Sensor was set 1.00 m, 2.00 m and 3.00 m. There is an example of detecting based on 2-Dimensional figure shown in

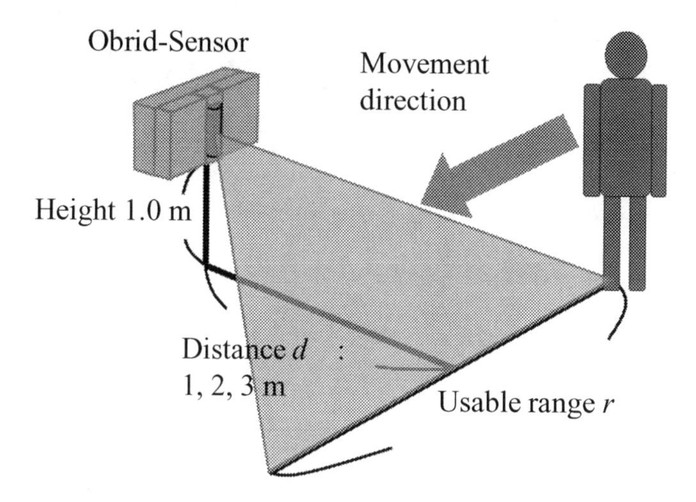

Fig. 4 Detection by Obrid-Sensor

Fig. 5. Figure 5 (a) shows the actual camera image; (b) shows the brightness distribution of (a). Line (A) shows the brightness distribution of background fetched in the first step. Line (B) is the brightness distribution with target person. The difference between (A) and (B) is shown as (C). It is clear that the position of target person can be detected according to the difference of brightness distribution.

To confirm effectiveness of the proposed method in experiments, a male in 20's is set as the target person who uses the restroom. The localization and falling down detection by the proposed method was implemented. The experimental device using Obrid-Sensor was designed and applied in experiments. As shown in Fig. 6, it is the photograph of the device with its size. In our design, two Obrid-Sensors, which were set in vertical and horizontal directions, were utilized. They are employed for detecting brightness distribution in the two directions, respectively. A restroom with the alignment as shown in Fig. 7 is employed as experimental environment. The environment can be considered as the most common restroom environment in normal life. The Obrid-Sensor is set in the right hand not far from entrance of the restroom. In the experiment, the target person was designed to move from entrance to toilet. The localization is implemented in the whole process. In order to confirm the usefulness of proposed method in localization, the localization of target person using brightness distribution are summarized as shown in Fig. 8, Fig. 9, Fig. 10 and Fig. 11. Figure 8 shows the condition when target person in the entrance of restroom and Fig. 9 shows the condition when the target person stands in the middle of the

(a) Camera image

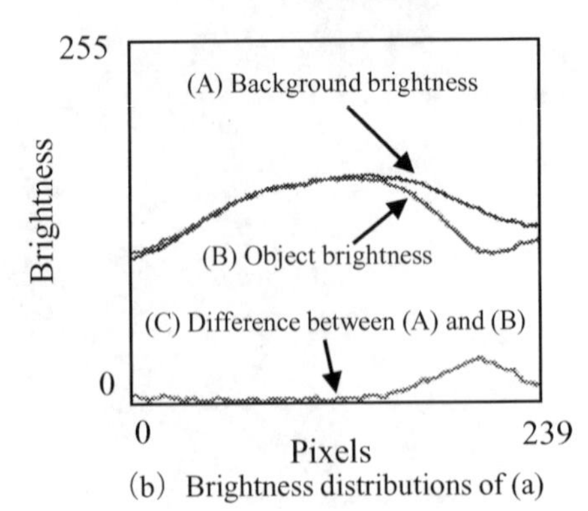

(b) Brightness distributions of (a)

Fig. 5 Example of detecting by Obrid-Sensor

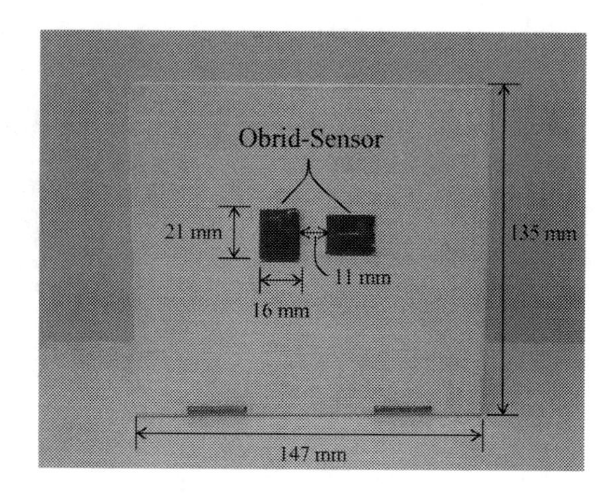

Fig. 6 Experimental device

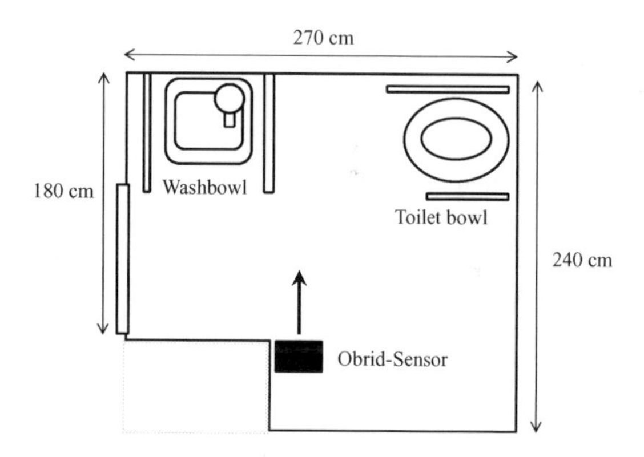

Fig. 7 Experimental environment in restroom

room. The condition of using toilet is shown in Fig. 10. The falling down results are shown in Fig. 11. The photographs in (a)s of the figures show the actual locations and gestures of the target person in experiments. As we see in (b)s of the figures, the peaks of brightness distribution varied according to the variation of distance between the target person and the sensor. The highest peak is 80 when the target person was in the middle of the restroom, nearest from the sensor. The lowest peak was 22, when the target person was at the entrance of the restroom. As we track the position of the peak in horizontal direction, it is clear that the target person can be localized by our definition: in entrance area, in middle of restroom or toilet using area. Meanwhile, gestures of the target person were also sensed in experiment. As we see in the (c)s in Fig. 8, Fig. 9, Fig. 10 and Fig. 11, gestures of the target person, standing, sitting and laying can be classified by defining proper threshold value and estimating centre of gravity. In the experiments, the states of target person were

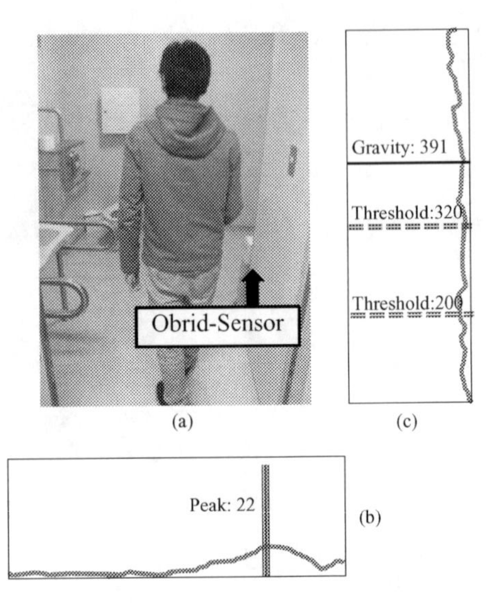

Fig. 8 Experimental results in the entrance of restroom

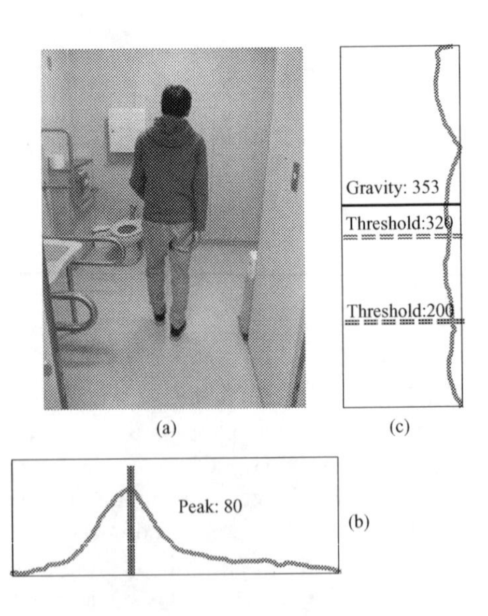

Fig. 9 Experimental results in the middle of restroom

defined by standing, sitting and lying. The states are divided by two thresholds. As we see in (c)s from Fig. 8 to Fig. 11, the thresholds were 320 and 200. If the gravity of target person was higher than 320, the state can be recognized as standing. If the gravity was between the two thresholds, the target person is recognized as sitting. The state with the gravity below 200 is defined as lying (falling down). According to the experimental results, it can be stated that the proposed method is effective in localization and falling down detecting in restroom with privacy protection.

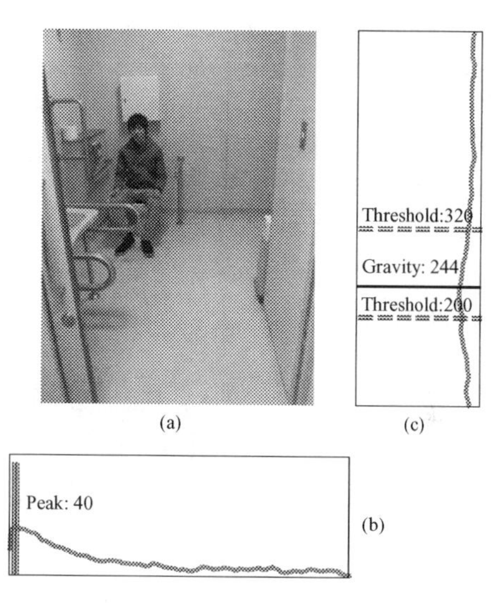

Fig. 10 Experimental results in use of toilet

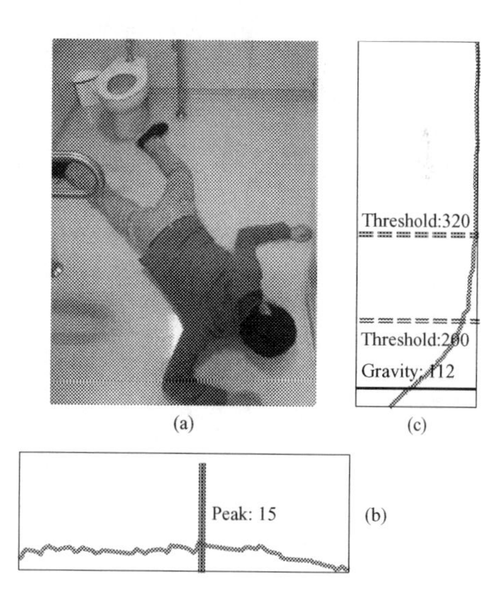

Fig. 11 Experimental results of falling down in restroom

6 Conclusions

Based on realization of the proposed privacy preserving sensor, the localization and falling down detection system in restroom is proposed in this research. According to brightness distribution detected by the sensor, localization and falling down detection functions can be utilized in restroom sensing without privacy offending. In order to validate effectiveness of the proposed method, a group of experiments have been implemented. In the experiments, it is confirmed that target person's position and movement status can be detected correctly owing to the brightness distribution by employing the proposed method. Therefore, this method is considered as an excellent choice for safety ensuring system in medical and welfare field.

Acknowledgements. This work was supported by JSPS KAKENHI Grant Number: 24700198.

References

1. Kitahara, I., Kogure, K., Hagita, N.: Stealth Vision: A method for video capturing system with protecting privacy. IEICE Technical Report 103, 89–94 (2004)
2. Koshimizu, T., Toriyama, T., Nishino, S., Babaguchi, N., Hagita, N.: Visual abstraction for privacy preserving video surveillance. IEICE Technical Report 105, 259–302 (2005)
3. Yabuta, K., Kitazawa, H., Tanaka, T.: A fixed monitoring camera image processing method satisfying both privacy protection and object recognition. IEICE Technical Report 105, 13–18 (2005)
4. Takenouchi, T., Morimoto, M., Kawahara, H., Takahashi, M., Yokota, H.: High-reliability and compact passive infrared detectors monitoring two independent areas. Matsushita Elect. Works Tech. Rep. 52, 62–68 (2004)
5. Mitsutoshi, S., Namba, Y., Serikawa, S.: Extraction of values representing human features in order to develop a privacy-preserving sensor. International Journal of ICIC 4, 883–895 (2008)
6. Hagiwara, M.: Neuro Fuzzy Genetic Algorithm, Sangyo Tosho (1994)
7. Serikawa, S., Simomura, T.: Proposal of a system of function discovery using a bug type of artifical life. IEEJ Trans. on Electronics, Information and Systems 118-C, 170–179 (1998)
8. Serikawa, S., Chhetri, B.B., Simomura, T.: Improvement of the search ability of S-System (a function discovery system). IEEJ Trans. on Electronics, Information and Systems 120-C, 1281–1282 (2000)
9. Shibata, J., Okuhara, K., Ishii, H.: Adaptive worker's arrangement and workload control for projectmanagement by genetic algorithm. International Journal of Innovative Computing, Information and Control 3, 175–188 (2007)
10. Koza, J.: Genetic Programming, Auto Discovery of Reusable Subprograms. MIT Press (1994)

An Approximate Reasoning Model for Medical Diagnosis

Irosh Fernando, Frans Henskens, and Martin Cohen

Abstract. Medical diagnosis is a classical example of approximate reasoning, and also one of the earliest applications of expert systems. The existing approaches to approximate reasoning in medical diagnosis are mainly based on Probability Theory and/or Multivalued Logic. Unfortunately, most of these approaches have not been able to model medical diagnostic reasoning sufficiently, or in a clinically intuitive way. The model described in this paper attempts to overcome the main limitations of the existing approaches.

Keywords: approximate reasoning, medical expert systems, inference model, psychiatry.

1 Introduction

Medical diagnostic reasoning is a typical application of approximate reasoning, in which the knowledge and reasoning are often characterised by uncertainty, vagueness and partial information. Whilst there are several approaches to approximate reasoning in medicine, most of them adopt Bayes Theorem (e.g. [16], [4], [2]) and/or Fuzzy Logic (e.g. [3], [10], [17]). The other approaches include Dempster-Shaper Theory [6], [13], Certainty Factors Model [14], and Cohens inductive probabilities [5].

Irosh Fernando · Frans Henskens
School of Electrical Engineering & Computer Science, University of Newcastle, Callaghan, NSW 2308, Australia
e-mail: irosh.fernando@uon.edu.au,
frans.henskens@newcastle.edu.au

Martin Cohen
The Mater Hospital, Hunter New England Area Health Service, Waratah, NSW 2298, Australia
e-mail: martin.cohen@hnehealth.nsw.gov.au

R. Lee (Ed.): *SNPD*, SCI 492, pp. 11–24.
DOI: 10.1007/978-3-319-00738-0_2

The approaches based on Bayes theorem have two main limitations: conformity of the total probabilities to be equal to one, and the assumption of conditional independence. Unfortunately these restrictions are not compatible with the real world expert clinical reasoning. For example, whilst $Probability(Disease = asthma) = 0.8$ and $Probability(Disease = bronchitis) = 0.9$ are clinically meaningful, it violates the axiom, which requires $Probability(Disease = asthma) + Probability(Disease = bronchitis) \leq 1$.

Fuzzy logic has been the most widely used multi-value logic, and in fact, it has been applied in early medical expert systems such as CADIAG-2 [1]. Unfortunately, the logical inference using Fuzzy Logic implication operators that are based on Fuzzy t-norm and t-conorm are less meaningful in medical diagnostic reasoning. Nonetheless, the Fuzzy Logic inference based on Sugeno-type Fuzzy inference [15], even though often used in contexts of control system applications, is able to provide a close approximation to medical diagnostic reasoning. The model proposed in this paper adopts a diagnostic inference similar to Sugeno-type (Takagi-Sugeno-Kang) Fuzzy inference applied in control system applications.

In real world clinical practice, symptoms are quantified using a number of symptom attributes (e.g. a symptom, chest pain, can be assessed quantitatively using attributes: duration, onset, severity of the pain). This quantification is important because the likelihood of a given diagnosis tends to vary according to the quantities of its related symptoms. The relationship between the likelihood of a diagnosis and the quantities of its relevant symptoms is often highly subjective and implicit. Clinicians usually acquire the understanding of these relationships with years of experience. Importantly, this relationship can be either positive or negative (i.e. whilst the presence of a one symptom increases the likelihood of a particular diagnosis, it may also decrease the likelihood of another). During diagnostic reasoning, clinicians actively look for these positive and negative relationships in order to rule in and rule out diagnostic possibilities. Conventional approaches based on Probability theory, Fuzzy logic, Dempster-Shaper Theory and Cohens inductive probabilities are unable to model the negative functional relationships.

The Certainty Factor model is able to represent the positive and negative functional relationships as measure of belief (MB) and measure of disbelief (MD) respectively [14], and is more intuitive to clinicians, compared to other approaches. Nevertheless, it models the clinical reasoning only at a superficial level. For example, it does not model the criticalness of the diseases, and also does not model the variations of diagnostic likelihoods based on different degrees of a symptom severity.

The proposed model described in this paper, is not only able to model both positive and negative functional relationships, but is also able to model the variations of likelihood and the criticalness of diagnoses according to the severities of symptoms, in a clinically intuitive manner.

2 Combining Evidence

The nature of medical diagnostic reasoning typically involves inferring a quantity of a dependent entity (e.g. a degree of likelihood of having a particular diagnosis) based on a set of quantities of independent entities (e.g. set of symptoms, each with a degree of severity). This section introduces a model, that captures the functional relationship between a quantity of a dependent entity (i.e. dependent variable) with a set of quantities of independent entities (i.e. conjugates of independent variables).

In order to explain this operator, let us consider the following two sets of entities: $S = \{s_1, s_2, \ldots, s_n\}; D = \{d_1, d_2, \ldots, d_m\}$, where each entity can be quantified using the quantification function, Q, that maps each entity, x_i of the entity set, X, into the range $[0,1]$. The function Q can be formally defined as, $Q : X \rightarrow [0,1]$. Now, given a quantity of an entity, $Q(s_i)$, the corresponding quantity of a related entity, d_j is expressed using the notation, $Q(d_j|s_i)$. Similar to the notation of conditional probability, $Q(d_j|s_i)$ means the quantity of d_j is conditional upon the quantity, $Q(s_i)$. A key feature of this model is the approximation of the relationship between the independent variable, $Q(s_i)$, and the dependent variable, $Q(d_j|s_i)$, using a function F_ij that is determined according to expert clinical judgment. The approximation of this functional relationship is described in Figure 1 using a hypothetical example. The relationship between $Q(s_i)$ and $Q(d_j|s_i)$ can be expressed as $Q(s_i)\, THEN\, Q(d_j|s_i)$, or equivalently as $Q(s_i) \Rightarrow Q(d_j|s_i)$ using the implication operator, where $Q(d_j|s_i) = F_{ij}(Q(s_i))$.

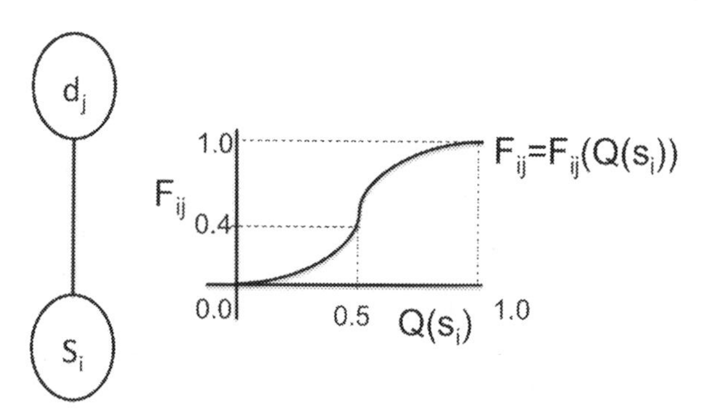

Fig. 1 $Q(d_j|s_i) = F_{ij}(Q(s_i))$

The above-described functional relationship is extended in order to combine several quantities of independent entities by introducing relative weights for each tuple, $< d_j, s_i >$. Weight, w_{ij} determines the relative importance of each s_i in relation to d_j. The relative importance of s_i in relation to d_j is expressed using the function, R which is expressed as:

$$R(s_i|d_j) = w_{ij} / \left(\sum_{i=1}^{n} w_{ij} \right) \tag{1}$$

Equation 1 is used in optimising the efficiency of the inference algorithm as described in the next section. Given a conjugate of independent variables, $Q(s_1) \wedge Q(s_2) \wedge \cdots \wedge Q(s_n)$, where all of them have either a positive or negative functional relationship with an entity, d_j, the corresponding quantity of d_j can be expressed and calculated as follows:

$IF\, Q(s_1) \wedge \cdots \wedge Q(s_n)\, THEN\, Q(d_j)\, WHERE\, Q(d_j) = Q(d_j|s_1 \wedge \cdots \wedge s_n)$ or $Q(s_1) \wedge \cdots \wedge Q(s_n) \Rightarrow Q(d_j)\, WHERE\, Q(d_j) = Q(d_j|s_1 \wedge \cdots \wedge s_n)$, where $Q(d_j|s_1 \wedge \cdots \wedge s_n)$ is calculated as follows :

$$Q(d_j|s_1 \wedge \cdots \wedge s_n) = \left(\sum_{i=1}^{n} F_{ij}(Q(s_i)).w_{ij} \right) / \left(\sum_{i=1}^{n} F_{ij}(Max(Q(s_i))).w_{ij} \right)$$

Since $F_{ij}(Max(Q(s_i))) = 1$, the above equation can be simplified as:

$$Q(d_j|s_1 \wedge \cdots \wedge s_n) = \left(\sum_{i=1}^{n} F_{ij}(Q(s_i)).w_{ij} \right) / \left(\sum_{i=1}^{n} w_{ij} \right) \quad (2)$$

When the equation 2 is applied to calculate $Q(d_j|s_i)$ using a single independent variable, $Q(s_i)$, it becomes evident that $Q(d_j|s_i) = F_{ij}(Q(s_i))$:

$$Q(d_j|s_i) = \frac{F_{ij}(Q(s_i))w_{ij}}{w_{ij}} = F_{ij}(Q(s_i))$$

Therefore, the above equation for calculating $Q(d_j|s_i \wedge \cdots \wedge s_n)$ can also be written as

$$Q(d_j|s_1 \wedge \cdots \wedge s_n) = \left(\sum_{i=1}^{n} Q(d_j|s_i).w_{ij} \right) / \left(\sum_{i=1}^{n} w_{ij} \right) \quad (3)$$

A similar equation is used in the Sugeno-type Fuzzy inference that has been described in control system applications [15]. In the rest of the paper, Equation 3 and the notation, $Q(d_j|s_i)$ will be used instead of $F_{ij}(Q(s_i))$ for convenience.

The above concepts are further elaborated using an example described in Figure 2. Given a rule, $Q(s_1) \wedge Q(s_2) \implies Q(d_1|s_1 \wedge s_2)$, this example describes how $Q(d_1|s_1 \wedge s_2)$ is calculated using $Q(s_1)$ and $Q(s_2)$. Both $Q(s_1)$ and $Q(s_2)$ have a positive functional relationship with the $Q(d_1|s_1)$ and $Q(d_1|s_2)$ respectively as described in Figure 2.

There can be a functional relationship between a dependent variable and a set of independent variables, where some of the independent variables have a positive functional relationship individually with the dependent variable while the rest have a negative functional relationship individually with the dependent variable. For example, let $(d_j|s_1^+ \wedge \cdots \wedge s_n^+ \wedge s_1^- \wedge \cdots \wedge s_m^-)$ be the dependent variable, where $Q(s_1^+),\ldots,Q(s_n^+)$ have positive functional relationships with $Q(d_j|s_1^+),\ldots,Q(d_j|s_n^+)$ respectively, whereas $Q(s_1^-),\ldots,Q(s_m^-)$ have negative functional relationships with $Q(d_j|s_1^-),\ldots,Q(d_j|s_n^-)$ respectively. $Q(d_j|s_1^+ \wedge \cdots \wedge s_n^+)$ and $Q(d_j|s_1^- \wedge \cdots \wedge s_m^-)$ can be calculated separately using Equation 3. Then, $Q(d_j|s_1^+ \wedge \cdots \wedge s_n^+ \wedge s_1^- \wedge \cdots \wedge s_m^-)$ can be calculated by adding $Q(d_j|s_1^+ \wedge \cdots \wedge s_n^+)$ and $Q(d_j|s_1^- \wedge \cdots \wedge s_m^-)$ as follows:

$$Q(d_j|s_1^+ \wedge \cdots \wedge s_n^+ \wedge s_1^- \wedge \cdots \wedge s_m^-) = Q(d_j|s_1^+ \wedge \cdots \wedge s_n^+) + Q(d_j|s_1^- \wedge \cdots \wedge s_m^-) \quad (4)$$

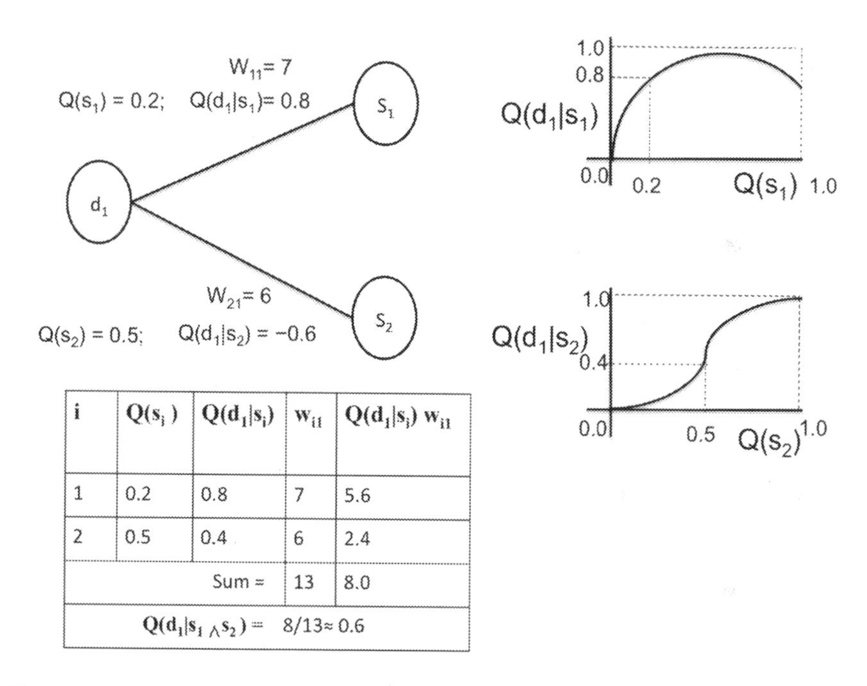

i	Q(s_i)	Q(d₁|s_i)	w_{i1}	Q(d₁|s_i) w_{i1}
1	0.2	0.8	7	5.6
2	0.5	0.4	6	2.4
		Sum =	13	8.0
	Q(d₁|s₁ ∧ s₂) =	8/13≈ 0.6		

Fig. 2 Calculating $Q(d_1|s_1 \wedge s_2)$, where $Q(s_1) \wedge Q(s_2) \Rightarrow Q(d_1|s_1 \wedge s_2)$

Extending the above example, Figure 3 describes how $Q(d_1|s_3 \wedge s_4)$ is calculated given $Q(s_3) \wedge Q(s_4) \Rightarrow Q(d_1|s_3 \wedge s_4)$, where the quantities, $Q(s_3)$ and $Q(s_4)$ have a negative relationship with the corresponding quantity of d_1. Given, $Q(d_1|s_1 \wedge s_2) = 0.6$ and $Q(d_1|s_3) \wedge s_4) = -0.4$, Equation 4 is used to calculate $Q(d_1|s_1 \wedge s_2 \wedge s_3 \wedge s_4)$ as follows:

$$Q(d_1|s_1 \wedge s_2 \wedge s_3 \wedge s_4) = Q(d_1|s_1 \wedge s_2) + Q(d_1|s_3 \wedge s_4) = 0.6 + -0.4 = 0.2$$

It is important to note that the quantity function, Q can have different meanings depending on the context. For example, $Q(d_j)$ can be interpreted as the likelihood of having the diagnosis, d_j or the degree of the severity of the diagnosis, d_j. A similar model with different notations has been previously introduced for diagnostic reasoning in psychiatry [8], [9], where $Q(d_j|s_1 \wedge \cdots \wedge s_n)$ can be calculated as follows:

$$Q(d_j) = Q(d_j|s_1 \wedge \cdots \wedge s_n) = \frac{1}{n} \sum_{i=1}^{n} Q(d_j|s_i) w_{ij}$$

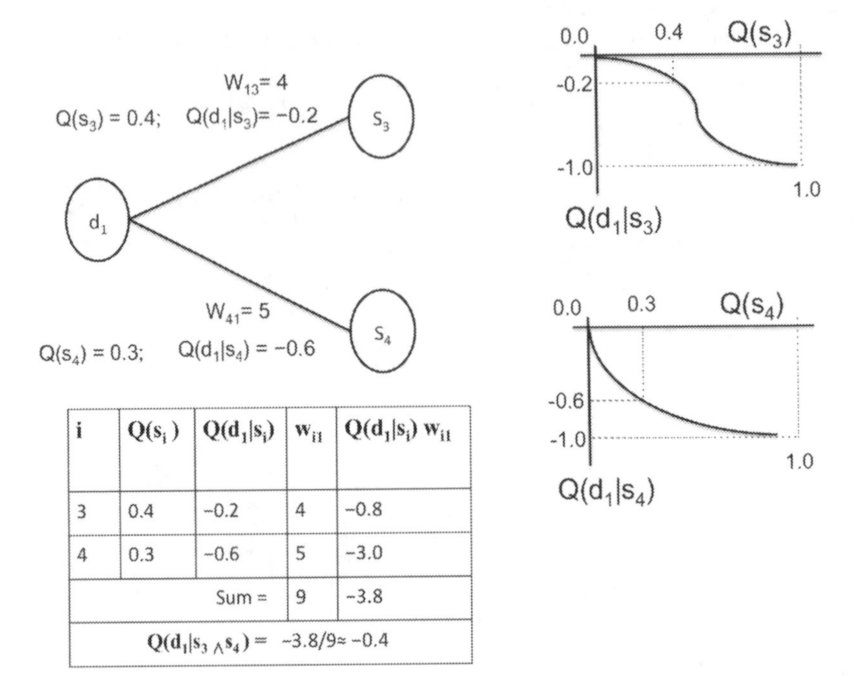

i	$Q(s_i)$	$Q(d_1\|s_i)$	w_{i1}	$Q(d_1\|s_i)\ w_{i1}$
3	0.4	−0.2	4	−0.8
4	0.3	−0.6	5	−3.0
		Sum =	9	−3.8
$Q(d_1\|s_3 \wedge s_4) =$ −3.8/9≈ −0.4				

Fig. 3 Calculating $Q(d_1|s_1 \wedge s_2)$, where $Q(s_1) \wedge Q(s_2) \Rightarrow Q(d_1|s_1 \wedge s_2)$

The authors' experimentation with clinical scenarios has indicated that the model described in Equation 2 approximates clinical reasoning more accurately than the above equation.

3 Diagnostic Inference

Using the above-described model for combining evidence, this section describes an example of diagnostic inference that uses a knowledgebase model consisting of three sets of entities: symptom attributes, symptoms and diagnoses, as described in Fig. 4.

The symptoms, which have positive functional relationships with a given diagnosis, d_j is called the inclusion criteria of d_j, whereas the symptoms, that have negative functional relationship are called the exclusion criteria of d_j.

Given a conjugate of independent variables, $Q(s_1) \wedge Q(s_2) \wedge \cdots \wedge Q(s_n)$, two kinds of functional relationships are defined using Equation 3 : $Q_L(d_j|s_1 \wedge \cdots \wedge s_n)$ and $Q_C(d_j|s_1 \wedge \cdots \wedge s_n)$, which are the likelihood and criticalness of d_j respectively.

Let us create a set of entities and a few inference rules for a sample knowledgebase.

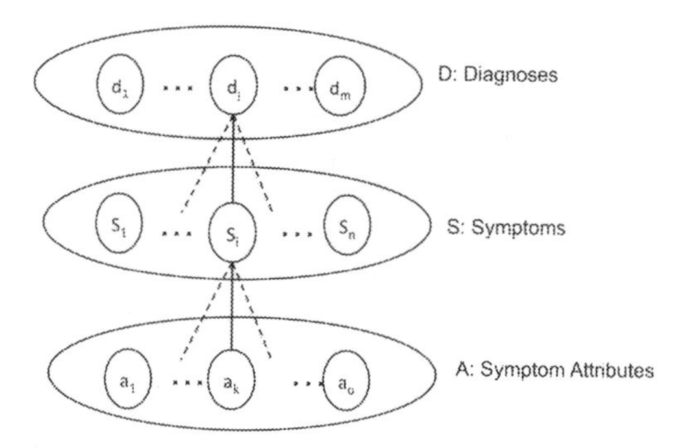

Fig. 4 Knowledgebase model

Diagnoses
d_1 =Depression; d_2=Generalised Anxiety
Symptoms s_1 =Loss of interest; s_2 = Fatigue; s_3 = Worrying thoughts ; s_4 =Suicidal thoughts
Attributes
a_1 =Duration ; a_2 =Number of activities given up
Rulebase
$Q(a_1) \wedge Q(a_2) \Rightarrow Q(s_1)$ where $Q(s_1) = Q(s_1|a_1 \wedge a_2)$;

$Q(s_1) \wedge Q(s_2) \wedge Q(s_3) \wedge Q(s_4) \wedge Q(s_5) \Rightarrow Q_L(d_1)\, WITH\, Q_C(d_1)\, WHERE\, Q_L(d_1) = Q_L(d_1|s_1 \wedge s_2 \wedge s_3 \wedge s_4 \wedge s_5)$, $Q_C(d_1) = Q_C(d_1|s_1 \wedge s_2 \wedge s_3 \wedge s_4 \wedge s_5)$;

$Q(s_1) \wedge Q(s_2) \wedge Q(s_3) \wedge Q(s_4) \Rightarrow Q_L(d_2)\ WITH\ Q_C(d_2)\ WHERE\ Q_L(d_2) = Q_L(d_2|s_1 \wedge s_2 \wedge s_3 \wedge s_4)$, $Q_c(d_2) = Q_c(d_2|s_1 \wedge s_2 \wedge s_3 \wedge s_4)$;

The sample knowledgebase created above is described as a graph in Figure 5, where the symptom, s_5 that is the only symptom in the exclusion criteria for the diagnosis, d_1 is connected to d_1 with a dotted line.
The inference model adopts the ST model introduced by Ramoni et al [12]. The ST model (described in Figure 6) is based on the logical inferences, abduction, deduction and induction, which were described by Charles Peirce [11]. Each step of the model is elaborated in the following sections.

4 Abstraction

Suppose the patient first reports the symptoms, s_1 and s_2. Abstraction involves, for example, in the case of s_1, mapping what the patient reported in his/her own words (e.g. Dont feel like doing anything) into the symptom, s_1 = Loss of interest, and quantifying it via eliciting the symptom attributes. The functional relationship between symptoms, s_1 and s_2, and the attributes, a_1 and a_2, are described in

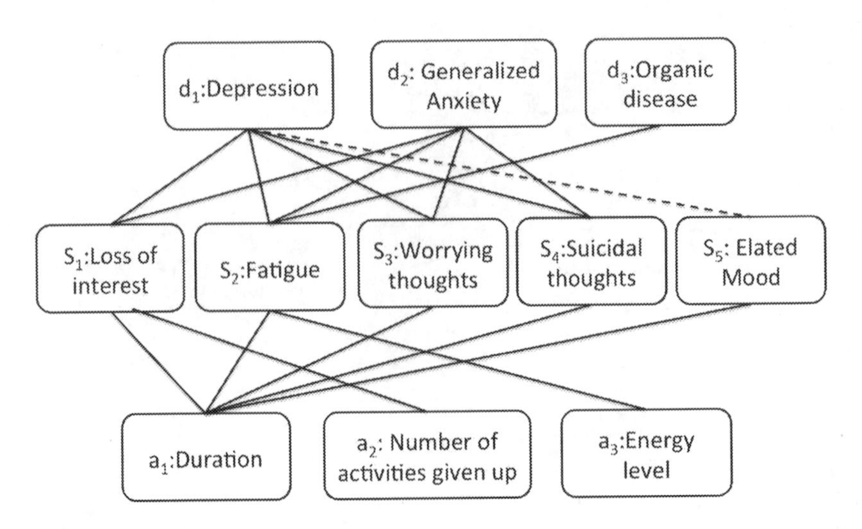

Fig. 5 Sample knowledgebase

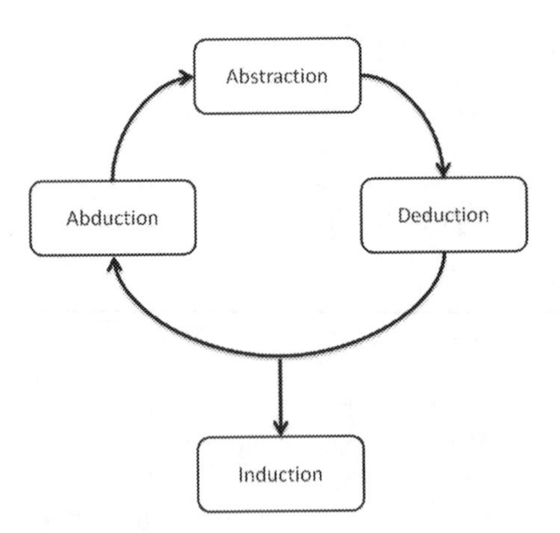

Fig. 6 ST model

Figure 7. Given, $Q(a_1) = 2$ and $Q(a_2) = 4$, the corresponding values, $Q(s_1|a_1) = 0.8$ and $Q(s_1|a_2) = 0.9$ are determined respectively, using the functional relationships described in Figure 7. Then, $Q(s_1|a_1 \wedge a_2)$ is calculated as described in the table in Figure 7, with the result that $Q(s_1) = Q(s_1|a_1 \wedge a_2) = 0.9$.

In a similar way, the symptom, s_2 is also abstracted and quantified using $Q(a_1)$ and $Q(a_3)$; then, $Q(s_2) = Q(s_2|a_1 \wedge a_3)$ is calculated. In order to proceed with this example, let us assume this calculation resulted in $Q(s_2) = Q(s_2|a_1 \wedge a_3) = 0.5$.

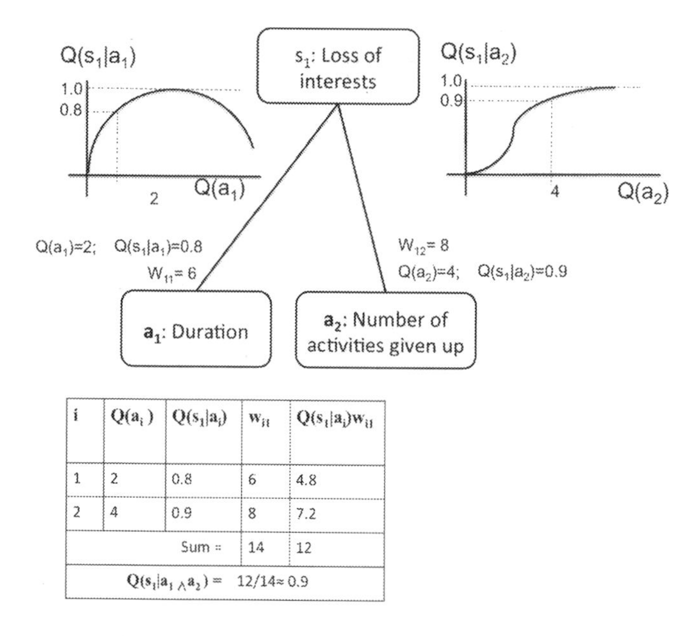

i	$Q(a_i)$	$Q(s_1 \mid a_i)$	w_{i1}	$Q(s_1 \mid a_i)w_{i1}$
1	2	0.8	6	4.8
2	4	0.9	8	7.2
		Sum =	14	12
		$Q(s_1 \mid a_1 \wedge a_2)$ =	12/14≈ 0.9	

Fig. 7 Abstraction of symptom, s_1

5 Abduction

The next step of the inference process is abduction. This involves hypothesising likely diagnoses and calculating their likelihoods and criticalness using the quantifications of the symptoms already done in the stage of abstraction:

$$Q(s_1) = Q(s_1|a_1 \wedge a_2) = 0.9$$
$$Q(s_2) = Q(s_2|a_1 \wedge a_3) = 0.5$$

As described in Figure 5, both s_1 and s_2 are connected to d_1 as well as d_2. In addition, s_2 is also connected to d_3. In many real world situations, any given set of symptoms can be connected to a large number of diagnoses indicating a large number of diagnostic possibilities, of which only some are relevant. In order to narrow down the list of likely diagnoses a threshold value, t_s is used. Only those symptoms with their quantities above the threshold value are used for determining the likely diagnoses.

Let $t_s = 0.5$. In this case, only $Q(s_1)$ is considered since $Q(s_2) \not> t_s$. Therefore we narrow down the search space by eliminating d_3 as we ignore s_2.

Now, using $Q(s_1) = 0.9$, we determine $Q_L(d_1|s_1) = 1.0$, $Q_L(d_2|s_1) = 0.6$ and $Q_C(d_1|s_1) = 0.2$, $Q_C(d_2|s_1) = 0.2$ as described in Figure 8. In real world situations, more than one symptom may indicate the same diagnosis. For example, suppose s_3 has been quantified and results in $Q(s_3) = 0.7$, which is above the threshold value.

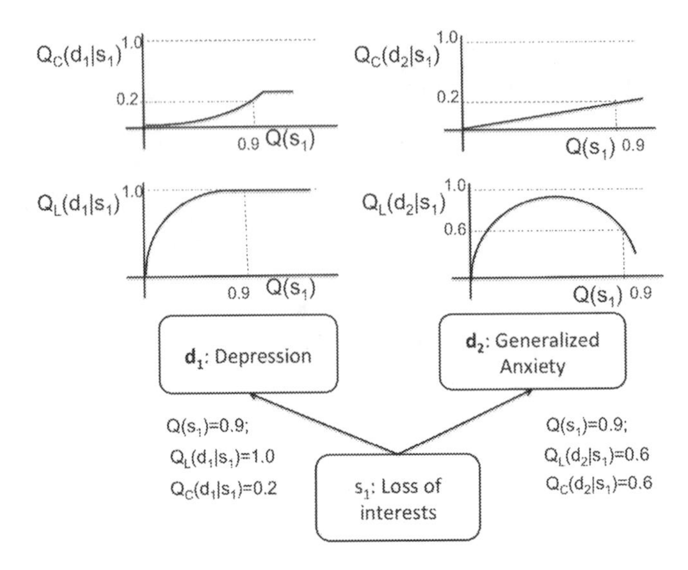

Fig. 8 Relationship between the quantity of the symptom, s_1 and the likelihood and criticalness of diagnoses d_1 and d_2.

In this case, both s_1 and s_3 are connected to d_1 (see Figure 5). Therefore, this involves calculating $Q_L(d_1|s_1 \wedge s_3)$ in a similar way as described in Figure 2.

6 Deduction

Deduction involves eliciting the rest of the symptoms associated with each of the likely diagnoses considered and quantified in the previous stage. The order of the symptoms that need to be quantified is determined according to the priority of each symptom, calculated using Equation 1. When an inference rule consists of both inclusion criteria and exclusion criteria, the symptoms in the exclusion criteria are quantified first. This is because, if the quantified symptoms in the exclusion criteria result in a high negative likelihood value, quantification of the symptoms in inclusion criteria can be terminated, thereby improving the efficiency of the inference algorithm.

In this example, s_5 is the only symptom in the exclusion criteria in relation to the diagnosis, d_1(elated mood contradicts the diagnosis of depression). Assuming that s_5 is abstracted and quantified with the result $Q(s_5) = 0.2$, Figure 9 describes how $Q_L(d_1|s_5)$ is derived using its functional relationship with $Q(s_5)$. Next, the remaining symptoms in the inclusion criteria are explored. In this example, this involves eliciting the rest of the symptoms for d_1 and d_2. In relation to both d_1 and d_2, the symptoms, s_3 and s_4 have to be elicited from the patient (see Figure 5).

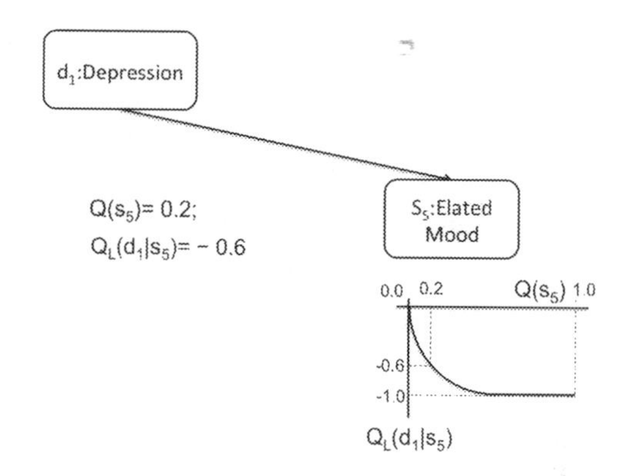

Fig. 9 Relationship between the quantity of the symptom, s_5 and the likelihood and criticalness of diagnoses d_1

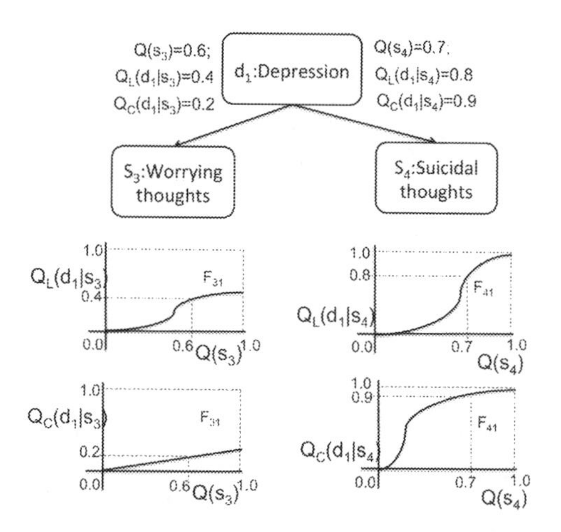

Fig. 10 Relationship between the quantity of the symptoms, s_3 and s_4, and the likelihood and criticalness of diagnosis d_1

It is possible that the quantified symptoms may indicate a large number of diagnoses including less relevant ones. In order to narrow down the search space further, another threshold value, t_d is used at this stage.

For example, let $t_d = 0.4$. In this case none of the diagnoses are eliminated since $Q(d_1|s_1) = 1.0 > t_d$ and also $Q(d_2|s_1) = 0.6 > t_d$.

The quantification of the symptoms, s_3 and s_4 are similar to what is described in Figure 7, therefore the details are omitted. Assuming this quantification resulted in $Q(s_3) = 0.6$ and $Q(s_4) = 0.7$, Figure 10 describes how $Q(d_1|s_3)$ and $Q(d_1|s_4)$ are determined. Similarly, $Q(d_2|s_3)$ and $Q(d_2|s_4)$ are also determined.

7 Induction

Induction involves testing the hypotheses (i.e. diagnostic likelihoods) by calculating the likelihood of the diagnoses considered in the previous stages and their criticalness. In our example, this involves calculating $Q_L(d_1)$, $Q_C(d_1)$, $Q_L(d_2)$, and $Q_C(d_2)$. Figure 11 describes how $Q_L(d_1|s_1 \wedge s_2 \wedge s_3 \wedge s_4)$ and $Q_L(d_1|s_5)$ are calculated. $Q_L(d_1|s_1 \wedge s_2 \wedge s_3 \wedge s_4)$ is the diagnostic likelihood (positive) of d_1 based on the symptoms in the inclusion criteria, whereas $Q_L(d_1|s_5)$ is the diagnostic likelihood (negative) of d_1 based on the symptom, s_5 in the exclusion criteria. Now, $Q_L(d_1|s_1 \wedge s_2 \wedge s_3 \wedge s_4 \wedge s_5)$ is calculated using Equation 4 as follows:

$$Q_L(d_1) = Q_L(d_1|s_1 \wedge s_2 \wedge s_3 \wedge s_4 \wedge s_5) = Q_L(d_1|s_1 \wedge s_2 \wedge s_3 \wedge s_4) + Q_L(d_1|s_5) = 0.7 - 0.6 = 0.1$$

Since $Q_L(d_1) = 0.1$ is less than the threshold value $t_d = 0.4$, the possibility of having the diagnosis (i.e. diagnostic hypothesis) d_1, is rejected.
In a similar way, $Q_C(d_1)$, $Q_L(d_2)$, and $Q_C(d_2)$ can also be calculated. If induction results in a large number of diagnoses, the threshold value, t_d can be used to filter out less likely diagnostic possibilities.

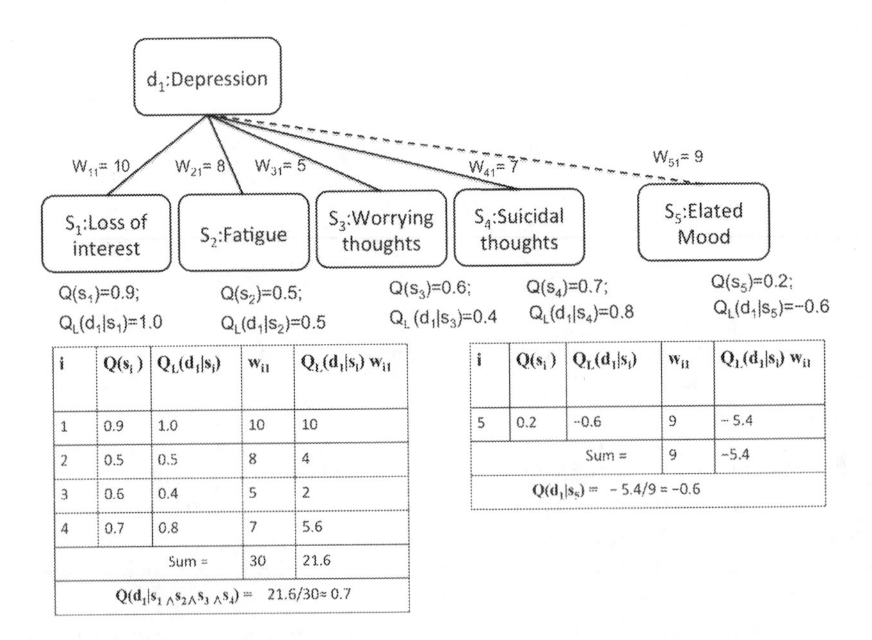

i	$Q(s_i)$	$Q_L(d_1\|s_i)$	w_{i1}	$Q_L(d_1\|s_i)\, w_{i1}$
1	0.9	1.0	10	10
2	0.5	0.5	8	4
3	0.6	0.4	5	2
4	0.7	0.8	7	5.6
		Sum =	30	21.6
$Q(d_1\|s_1 \wedge s_2 \wedge s_3 \wedge s_4)$ = 21.6/30≈ 0.7				

i	$Q(s_i)$	$Q_L(d_1\|s_i)$	w_{i1}	$Q_L(d_1\|s_i)\, w_{i1}$
5	0.2	-0.6	9	-5.4
		Sum =	9	-5.4
$Q(d_1\|s_5)$ = -5.4/9 = -0.6				

Fig. 11 Deriving the likelihood of the diagnosis, d_1 using the symptoms in the inclusion criteria and exclusion criteria

8 Conclusion

This paper describes a model that approximates reasoning in medical diagnosis by adopting the ST model and a Sugeno-type Fuzzy inference. Unlike some of the previous approaches, the proposed approach is able to model both negative and positive functional relationships, and also the likelihood and criticalness of diagnoses that vary according the severity of symptoms.
Furthermore, the proposed model is not only simple for clinicians with non-technical background to understand, but also intuitive to clinicians compared to other models. Therefore it is able to overcome the main limitations of other approaches that have been described earlier.

Two major challenges in medical expert system development include difficulties implementing a sufficiently large knowledgebase and engaging clinicians. The simplicity of the model and its clinical intuitiveness offers the advantage of being able to engage clinicians and actively involve them in a collaborative development of the knowledgebase. A prototype of a web-based platform for collaborative development of a knowledgebase, and diagnostic decision support in psychiatry, has been implemented using this model [7].

References

1. Adlassnig, K.P., Kolarzs, G.: Representation and semiautomatic acquisition of medical knowledge in cadlag-1 and cadiag-2. Computers and Biomedical Research 19, 63–79 (1986)
2. Andreassen, S., Jensen, F.V., Olesen, K.G.: Medical expert systems based on causal probabilistic networks. International Journal of Bio-Medical Computing 28, 1–30 (1991)
3. Boegl, K., Adlassnig, K.P., Hayashi, Y., Rothenfluh, T.E., Leitich, H.: Knowledge acquisition in the fuzzy knowledge representation framework of a medical consultation system. Artificial Intelligence in Medicine 30, 1–26 (2004)
4. Chard, T., Rubenstein, E.M.: A model-based system to determine the relative value of different variables in a diagnostic system using bayes theorem. International Journal of Bio-Medical Computing 24, 133–142 (1989)
5. Cohen, L.J.: Applications of Inductive Logic. Oxford University Press, Clarendon (1980)
6. Dempster, A.: Upper and Lower Probabilities Induced by a Multivalued Mapping. In: Yager, R.R., Liu, L. (eds.) Classic Works of the Dempster-Shafer Theory of Belief Functions, vol. 219, pp. 57–72. Springer, Heidelberg (2008)
7. Fernando, I., Henskens, F.A.: A web-based flatform for collaborative development of a knowledgebase for psychiatric case formulation and treatment decision support. In: IADIS e-Health 2012 International Conference, Lisban, Portugal (2012)
8. Fernando, I., Henskens, F.A., Cohen, M.: A domain specific conceptual model for a medical expert system in psychiatry, and a development framework. In: IADIS e-Health 2011 International Conference, Rome, Italy (2011)
9. Fernando, I., Henskens, F.A., Cohen, M.: A domain specific expert system model for diagnostic consultation in psychiatry. In: 12th ACIS International Conference on Software Engineering, Artificial Intelligence, Networking and Parallel/Distributed Computing, SNPD 2011 (2011)

10. Godo, L., de Mántaras, R.L., Puyol-Gruart, J., Sierra, C.: Renoir, pneumon-ia and terap-ia: three medical applications based on fuzzy logic. Artificial Intelligence in Medicine 21, 153–162 (2001)
11. Peirce, C.S.: Illustrations of the logic of science, sixth paper-deduction, induction, hypothesis. The Popular Science Monthly 1, 470–482 (1878)
12. Ramoni, M., Stefanelli, M., Magnani, L., Barosi, G.: An epistemological framework for medical knowledge-based systems. IEEE Transactions on Systems, Man and Cybernetics 22, 1361–1375 (1992)
13. Shafer, G.: A Mathematical Theory of Evidence. Princeton University Press (1976)
14. Shortliffe, E.H., Buchanan, B.G.: A model of inexact reasoning in medicine. Mathematical Biosciences 23, 351–379 (1975)
15. Sugeno, M.: Industrial applications of fuzzy control. Elsevier Science (1985)
16. Todd, B.S., Stamper, R., Macpherson, P.: A probabilistic rule-based expert system. International Journal of Bio-Medical Computing 33, 129–148 (1993)
17. Vetterlein, T., Ciabattoni, A.: On the (fuzzy) logical content of cadiag-2. Fuzzy Sets and Systems 161, 1941–1958 (2010)

Study on Adiabatic Quantum Computation in Deutsch-Jozsa Problem

Shigeru Nakayama and Peng Gang

Abstract. Adiabatic quantum computation has been proposed as a quantum algorithm with adiabatic evolution to solve combinatorial optimization problem, then it has been applied to many problems like satisfiability problem. Among them, Deutsch and Deutsch-Jozsa problems have been tried to be solved by using adiabatic quantum computation. In this paper, we modify the adiabatic quantum computation and propose to solve Deutsch-Jozsa problem more efficiently by a method with higher observation probability.

1 Introduction

Adiabatic quantum computation[1, 2, 3] was proposed as a quantum algorithm with adiabatic evolution to solve combinatorial optimization problem. Then it has been applied to many problems like satisfiability problem (SAT)[4]. Among them, Deutsch-Jozsa problem[5, 6, 7] has been tried to be solved by using adiabatic quantum computation.

In this paper, we modify the adiabatic quantum computation and propose a method to solve Deutsch-Jozsa problem more efficiently with higher observation probability.

2 Deutsch-Jozsa Problem

Deutsch-Jozsa problem was developed by David Deutsch and Richard Jozsa in 1992. If a function $f(x)$ has a binary value for an input variable x of n bits as follows;

Shigeru Nakayama
Dept. of Information Science and Biomedical Engineering, Kagoshima University, Kagoshima City, 890-0065 Japan
e-mail: shignaka@ibe.kagoshima-u.ac.jp

Peng Gang
Dept. of Computer Science, Huizhou University, Huizhou City, China
e-mail: peng@hzu.edu.cn

R. Lee (Ed.): *SNPD*, SCI 492, pp. 25–35.
DOI: 10.1007/978-3-319-00738-0_3

$$f(x|x \in \{0,1\}^n) \to \{0,1\} \tag{1}$$

The problem is to minimize the inquiry to the function to decide whether it is constant (uniform) or balanced (equal). For example, we consider the case of the input variable x with $n = 2$ bit. When the function always has the same values for any input values of $x = 00, 01, 10, 11$, *i.e.*, $f(00) = f(01) = f(10) = f(11) = 0$ *or* 1, the function is called a constant function. On the other hand, when the function values $f(00)$, $f(01)$, $f(10)$, $f(11)$ have 0 twice and 1 twice, the function is called a balanced function. It is promised that the function should be constant or balanced. In the classic algorithm, we must inquire the function three times, like $f(00)$, $f(01)$, and $f(10)$. In general, classic algorithm requires $2^{n-1}+1$ queries to the function with n bit of the input valuable.

However, in Deutsch-Jozsa algorithm, only one inquiry to $f(x)$ is enough to determine whether this function is constant or balanced. In late years quantum algorithms such as adiabatic quantum calculation or quantum random walk are suggested and have been applied to NP complete problem. Deutsch-Jozsa algorithm with $n = 2$ needs three qubits, but adiabatic quantum algorithm needs only two qubits. Moreover, we propose more efficient method for the adiabatic quantum computation to solve Deutsch-Jozsa problem.

3 Adiabatic Quantum Computation

3.1 Discrete Expression of Schrödinger Equation

Deutsch-Jozsa problem is to find the state vector $|\psi(t)\rangle$ where the eigenvalue (energy) of time-independent Hamiltonian H is minimized by using the adiabatic quantum computation. Adiabatic quantum computation uses the Schrödinger equation.

$$i\hbar \frac{\partial |\psi(t)\rangle}{\partial t} = H|\psi(t)\rangle . \tag{2}$$

The continuous differential equation must be converted into discrete expression for computer simulation.

$$|\psi(t+\Delta)\rangle = e^{-iH\Delta}|\psi(t)\rangle , \tag{3}$$

where Plank constant is used as $\hbar = 1$ for simplicity, and Δ is a differential time contributing to a phase scaling parameter on the right hand side of eq. (3).

3.2 Hamiltonian and Unitary Transform

Adiabatic quantum computation introduces a step parameter s ranging from 0 to 1. Then, the Hamiltonian H can be written as

$$H(s) = (1-s)H_i + sH_f , \tag{4}$$

as a function of step parameter s, where H_i is Hamiltonian of the initial state independent of any problem and H_f is Hamiltonian of the final state depending on a given problem. When the step parameter s is gradually changed from 0 to 1, the Hamiltonian will be changed from the initial Hamiltonian to final Hamiltonian.

Therefore, the state vector $\left|\psi^{(h)}\right\rangle$ gradually changes with each step by unitary transform as follows;

$$\left|\psi^{(h)}\right\rangle = e^{-i(1-s)H_i\Delta - isH_f\Delta}\left|\psi^{(h-1)}\right\rangle \tag{5}$$

where an integer parameter hi$1 \leq h \leq j+1$jis calculated from $s = h/(j+1)$, j is the number of repetition steps, and $U^{(h)}$ is a unitary transform for adiabatic quantum evolution written as,

$$U^{(h)} = e^{-i(1-s)H_i\Delta - isH_f\Delta}. \tag{6}$$

The same parameters in this computation are used to compare with the theory of Das et al.[5]. For example, the number of repetition steps is also used as $j = 4$ in our numerical simulation, and Δ is a phase scaling parameter used as $\Delta = 1$ required for adiabatic quantum evolution.

$U_{SM}^{(h)} = e^{-i(1-s)H_i}$ has non-diagonal elements and is unitary transform for state mixing between interacting states. This causes diversification of the solutions. On the other hand, $U_{PS}^{(h)} = e^{-isH_f}$ has only diagonal elements and is unitary transform for phase shifting where the phase of a state with high cost is shifted very much and that of a state with low cost is not shifted much. This results in intensification for searching solutions.

3.3 *Hamiltonian Modified by Nonlinear Step Functions*

We modify the linear step parameter s to a nonlinear step function $p(s)$ changing from 0 to 1. Hamiltonian can be written as

$$H(s) = (1 - p(s))H_i + p(s)H_f, \tag{7}$$

where we introduce a nonlinear step function $p(s)$ changing from 0 to 1. The nonlinear step function must be monotonic to fulfill both $p(0) = 0$ and $p(1) = 1$. In the case of linear step function,

$$p(s) = s. \tag{8}$$

Although the step function $p(s)$ is originally used as a linear function[1, 5], we propose the nonlinear step function and did the computer simulation to solve Deutsch problem. Here, we propose the cubic step function,

$$p(s) = 4(s - 1/2)^3 + 1/2, \tag{9}$$

as shown in Fig. 1, and compare the linear step function with the cubic step function.

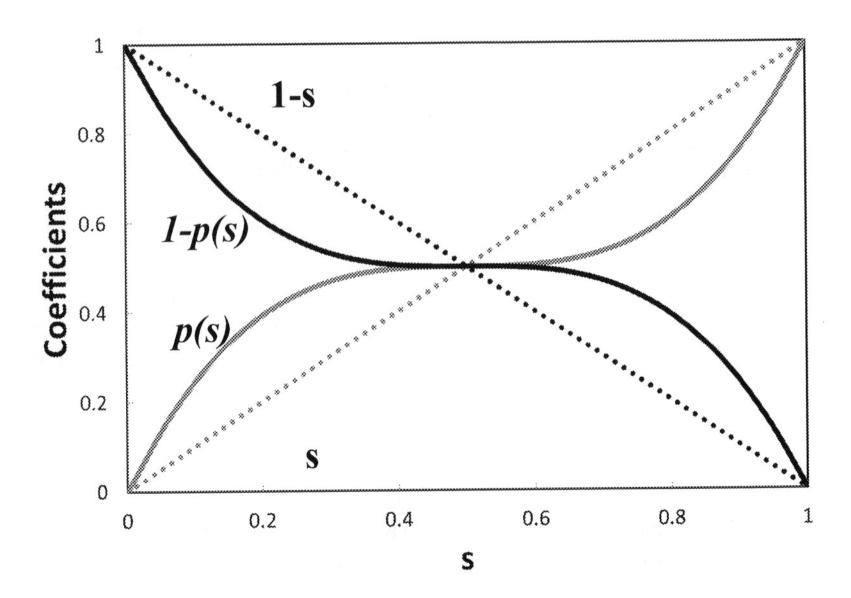

Fig. 1 Two coefficients in Hamiltonian with linear or cubic step function as a function of step parameter

4 Hamiltonian of Deutsch-Jozsa Problem

4.1 Initial Hamiltonian of Deutsch-Jozsa Problem

According to Das et al.[5], the Hamiltonian H_i of the initial state,

$$|\psi_i\rangle = \frac{1}{2}(|00\rangle + |01\rangle + |10\rangle + |11\rangle), \tag{10}$$

of two qubits $n = 2$ used in Deutsch-Jozsa problem is given by,

$$H_i = I - |\psi_i\rangle\langle\psi_i| = \frac{1}{4}\begin{pmatrix} 3 & -1 & -1 & -1 \\ -1 & 3 & -1 & -1 \\ -1 & -1 & 3 & -1 \\ -1 & -1 & -1 & 3 \end{pmatrix} \tag{11}$$

where I is a 4×4 unit matrix.

4.2 Final Hamiltonian of Deutsch-Jozsa Problem

The Hamiltonian H_f of the final state $|\psi_f\rangle$ is given by

$$H_f = I - |\psi_f\rangle\langle\psi_f|, \tag{12}$$

where

$$\left|\psi_f\right\rangle = \alpha\left|00\right\rangle + \frac{\beta}{\sqrt{3}}[\left|01\right\rangle + \left|10\right\rangle + \left|11\right\rangle], \tag{13}$$

depends on the function $f(x)$ given in Deutsch·Jozsa problem. The coefficients α and β are given by,

$$\alpha = \frac{1}{4}|(-1)^{f(00)} + (-1)^{f(01)} + (-1)^{f(10)} + (-1)^{f(11)}|, \tag{14}$$

$$\beta = 1 - \alpha, \tag{15}$$

respectively. If the function is a constant function such as $f(00) = f(01) = f(10) = f(11) = 0$ *or* 1, then $\alpha = 1, \beta = 0$. If the function is a balanced function such as $f(00) = f(01) = 0, f(10) = f(11) = 1$, then $\alpha = 0, \beta = 1$.

Using the properties of $\alpha + \beta = 1, \alpha\beta = 0$, the Hamiltonian H_f of the final state $\left|\psi_f\right\rangle$ is given by,

$$H_f = \frac{1}{3}\begin{pmatrix} 3\beta & 0 & 0 & 0 \\ 0 & 3-\beta & -\beta & -\beta \\ 0 & -\beta & 3-\beta & -\beta \\ 0 & -\beta & -\beta & 3-\beta \end{pmatrix} \tag{16}$$

5 Adiabatic Quantum Computation in Deutsch-Jozsa Problem

5.1 *Initial State in Deutsch-Jozsa Problem*

We prepare the initial state in Deutsch-Jozsa Problem as follows;

$$\left|\psi^{(0)}\right\rangle = \frac{1}{2}(\left|00\right\rangle + \left|01\right\rangle + \left|10\right\rangle + \left|11\right\rangle) = \frac{1}{2}\begin{pmatrix} 1 \\ 1 \\ 1 \\ 1 \end{pmatrix} \tag{17}$$

where equal probability amplitudes are provided to $\left|00\right\rangle$, $\left|01\right\rangle$, $\left|10\right\rangle$, and $\left|11\right\rangle$. The initial state must be independent of the property of the function given in Deutsch·Jozsa problem.

5.2 *Energy State of Deutsch-Jozsa Hamiltonian*

After this,we assume that the function in Deutsch-Jozsa problem is constant, then we use these values of $\alpha = 1$ and $\beta = 0$. By using the nonlinear step function $p(s)$, Hamiltonian ranging from Hamiltonian H_0 of the initial state $\left|\psi_0\right\rangle$ to Hamiltonian H_1 of the final state $\left|\psi_1\right\rangle$ is totally written as,

$$H(s) = \frac{1}{4}\begin{pmatrix} 3-3p(s) & p(s)-1 & p(s)-1 & p(s)-1 \\ p(s)-1 & 3+p(s) & p(s)-1 & p(s)-1 \\ p(s)-1 & p(s)-1 & 3+p(s) & p(s)-1 \\ p(s)-1 & p(s)-1 & p(s)-1 & 3+p(s) \end{pmatrix}. \tag{18}$$

By using the eigen equation of this Hamiltonian,

$$|H(s) - \lambda I| = 0, \tag{19}$$

we can obtain two energy eigen values $\lambda_{1,2,3,4}$ given by,

$$\lambda_{1,2,3,4} = 1, 1, \frac{1}{2}[1 \pm \sqrt{1 - 3p(s) + 3p(s)^2}]. \tag{20}$$

Figure 2 shows evolution of four energy levels of eigenvalues $\lambda_{1,2,3,4}$ as functions of step parameter s and step function $p(s)$ by adiabatic quantum computation in Deutsch·Jozsa problem. The highest eigenvalue of the Hamiltonian is 1, which is two-fold degenerate. As shown in Fig. 2, there are long flat lines for the nonlinear step function, which mean long interaction between the upper energy level and the lower energy level. The lasting of long interaction causes strong transition between the two levels. However, there are short flat lines for the linear step function, which have only short time for strong transition.

At the initial step $s = 0$, the four eigenstates of the Hamiltonian are written as,

$$\frac{-|00\rangle + |11\rangle}{\sqrt{2}}, \frac{-|00\rangle + |10\rangle}{\sqrt{2}}, \frac{-|00\rangle + |01\rangle}{\sqrt{2}}, \frac{|00\rangle + |01\rangle + |10\rangle + |11\rangle}{2}. \tag{21}$$

These eigenstates of the Hamiltonian at the final step $s = 1$ are given by,

$$|11\rangle, |10\rangle, |01\rangle, |00\rangle, \tag{22}$$

respectively. It is considered that the initial ground state at $s = 0$, $(|00\rangle + |01\rangle + |10\rangle + |11\rangle)/2$, evolves to the final ground state at $s = 1$, $|00\rangle$, after several unitary transforms.

5.3 Evolution of State Vector by Unitary Transform

We are assuming that the function of Deutsch-Jozsa problem is a constant function $f(0) = f(1)$, and we do the numerical calculation in the case of $\alpha = 1, \beta = 0$. The final state will be,

$$\left|\psi^{(9)}\right\rangle = \alpha|00\rangle + \frac{\beta}{\sqrt{3}}[|01\rangle + |10\rangle + |11\rangle] \approx |00\rangle \tag{23}$$

Therefore, we should make the observation probability of the state vector $|00\rangle$ as higher as possible than 25% in the initial state $\left|\psi^{(0)}\right\rangle = (|00\rangle + |01\rangle + |10\rangle + |11\rangle)/2$.

The initial state $\left|\psi^{(0)}\right\rangle$ with equal superposition state shown in Gauss space of Fig. 3(0), is evolved by applying the unitary transform $U^{(1)}$ with the cubic step

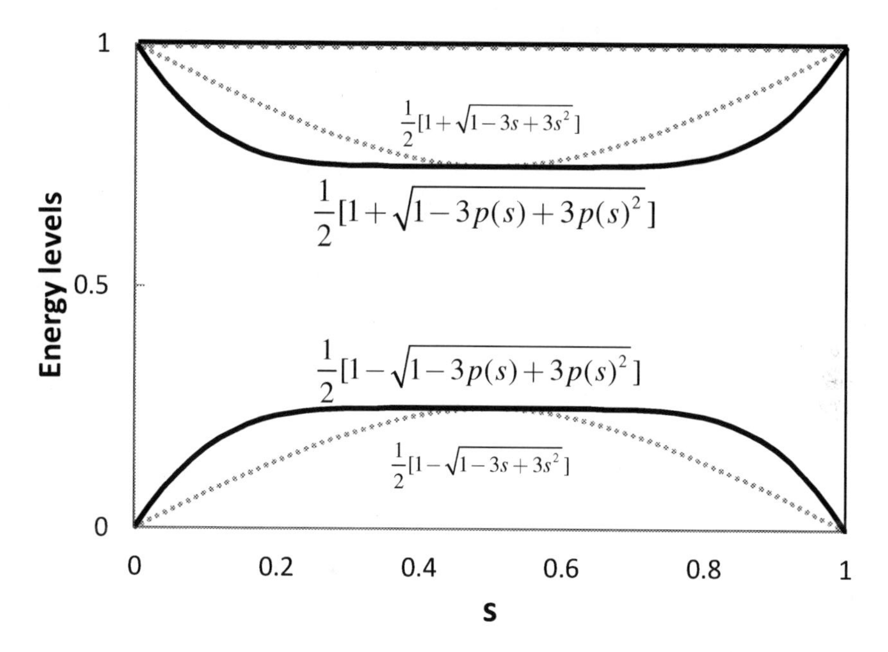

Fig. 2 Evolution of four energy levels by adiabatic quantum computation in Deutsch-Jozsa problem

function at the first step $s = 1/9(h = 1)$. The first evolved state vector can be calculated as follows;

$$\left|\psi^{(1)}\right\rangle = U^{(1)}\left|\psi^{(0)}\right\rangle \approx \begin{pmatrix} 0.5340-0.0117i \\ 0.4716-0.1259i \\ 0.4716-0.1259i \\ 0.4716-0.1259i \end{pmatrix} \begin{matrix} \leftarrow |00\rangle \\ \leftarrow |01\rangle \\ \leftarrow |10\rangle \\ \leftarrow |11\rangle \end{matrix} \tag{24}$$

where two state vectors change and the complex numbers come out. The two state vectors are shown in Gauss space of Fig. 3(1).

This unitary transform is repeated five times at each step. At the step $s = 8/9(h = 8)$, we obtain the final solution state shown in Fig. 3(8) as follows;

$$\left|\psi^{(8)}\right\rangle = U^{(8)}\left|\psi^{(7)}\right\rangle \approx \begin{pmatrix} -0.1520-0.9864i \\ 0.0239-0.0277i \\ 0.0239-0.0277i \\ 0.0239-0.0277i \end{pmatrix} \begin{matrix} \leftarrow |00\rangle \\ \leftarrow |01\rangle \\ \leftarrow |10\rangle \\ \leftarrow |11\rangle \end{matrix} \tag{25}$$

where the probability amplitude of observing the state vector $|00\rangle$ is getting high.

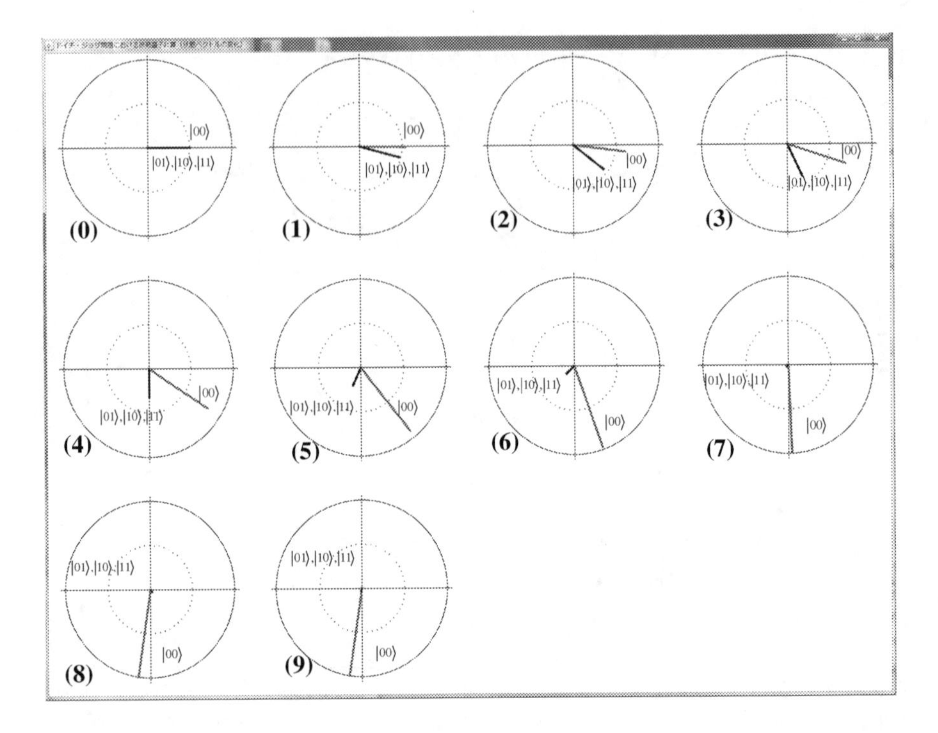

Fig. 3 Evolution of two state vectors by adiabatic quantum computation as a function of step parameter s in Gauss space

5.4 *Evolution of Observation Probability by Adiabatic Quantum Computation*

In adiabatic quantum computation, the observation probability of each state vector at each step can be calculated from taking the square of the absolute value of the probability amplitude. Then, the observation probability $P^{(1)}$ of the two state vectors at the first step $s = 1/9(h = 1)$ is given as follows.

$$P^{(1)} \approx \begin{pmatrix} |0.5340-0.0117i|^2 \\ |0.4716-0.1259i|^2 \\ |0.4716-0.1259i|^2 \\ |0.4716-0.1259i|^2 \end{pmatrix} \approx \begin{pmatrix} 0.2852 \\ 0.2383 \\ 0.2383 \\ 0.2383 \end{pmatrix} \begin{matrix} \leftarrow |00\rangle \\ \leftarrow |01\rangle \\ \leftarrow |10\rangle \\ \leftarrow |11\rangle \end{matrix} \tag{26}$$

The observation probability of the $|00\rangle$ state increases to some extent than that of the initial state, and becomes 28.52%. The observation probability of the other states decreases to 23.83%. Since the superposition state is normalized, the addition of two probabilities becomes 1 even though there is some fraction effect.

At the next step $s = 2/9(h = 2)$, the observation probability of the $|00\rangle$ state increases to 37.72% and that of the other states decreases to 20.76% as follows;

$$P^{(2)} \approx \begin{pmatrix} |0.6095 - 0.0760i|^2 \\ |0.3599 - 0.2794i|^2 \\ |0.3599 - 0.2794i|^2 \\ |0.3599 - 0.2794i|^2 \end{pmatrix} \approx \begin{pmatrix} 0.3772 \\ 0.2076 \\ 0.2076 \\ 0.2076 \end{pmatrix} \begin{matrix} \leftarrow |00\rangle \\ \leftarrow |01\rangle \\ \leftarrow |10\rangle \\ \leftarrow |11\rangle \end{matrix} \tag{27}$$

At the step $s = 8/9(h = 8)$, the observation probability of the $|00\rangle$ state increases to 99.6% and that of the other states decreases to 0.13% as follows;

$$P^{(8)} \approx \begin{pmatrix} |-0.1520 - 0.9864i|^2 \\ |0.0239 - 0.0277i|^2 \\ |0.0239 - 0.0277i|^2 \\ |0.0239 - 0.0277i|^2 \end{pmatrix} \approx \begin{pmatrix} 0.9960 \\ 0.0013 \\ 0.0013 \\ 0.0013 \end{pmatrix} \begin{matrix} \leftarrow |00\rangle \\ \leftarrow |01\rangle \\ \leftarrow |10\rangle \\ \leftarrow |11\rangle \end{matrix} \tag{28}$$

At the last step $s = 9/9(h = 9)$, the observation probability of the $|00\rangle$ state is the same as 99.6% as shown in Fig. 4. Since there is no state mixing at the last step $s = 1$, the observation probability will not change. As a result, we start with four equal probabilities of observing the states of $|00\rangle$, $|01\rangle$, $|10\rangle$, and $|11\rangle$ in the initial state, then we obtain the solution of $|00\rangle$ with high observation probability of 99.6% after the four unitary transforms in the adiabatic quantum computation.

On the other hand, we could not increase the observation probability in the linear step function $p(s) = s$. Then, we get the final observation probability 86.27% of finding the solution as shown in Fig. 4.

6 Considerations on Energy Gap in Adiabatic Quantum Computation

The energy gap between the two states shown in Fig. 2 indicates an interaction between the two levels. The bigger the energy gap, the stronger the interaction. The strong interaction results in fast population transitions between the levels, then we can achieve fast calculation.

However, if there is no energy gap, then no interaction exists between the levels. This is called level-crossing where adiabatic quantum computation does not work. Therefore, we must choose the step function to obtain a wider energy gap or keep a strong interaction for a long time. We can quickly reduce the observation probability amplitude of the non-solution state as low as possible, and increase that of the solution state as high as possible.

In Fig. 2, the energy gap between the ground state and the excited state becomes the smallest at $s = 0.5$ in the linear step function $p(s) = s$. At this point, the interaction is the strongest. On the other hand, the flat region continues at the same gap for a long time in the cubic step function $p(s) = 4(s - 1/2)^3 + 1/2$. Thus, strong interaction must be working for a long time and we can see that this causes observation probability increase very quickly.

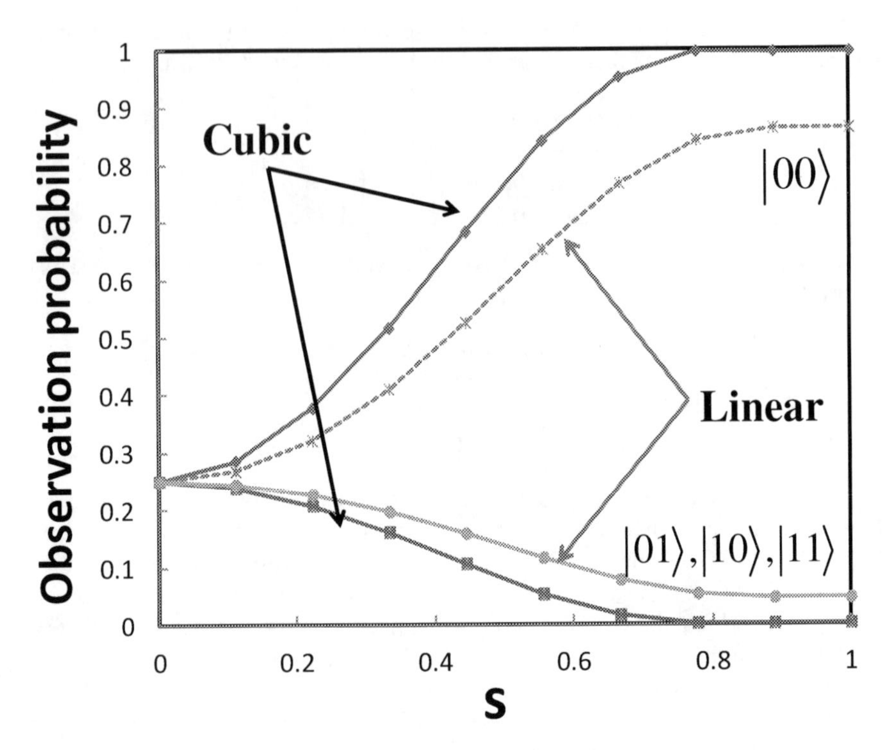

Fig. 4 Observation probabilities $P^{(h)}$ of two state vectors as a parameter of step parameter s

7 Conclusion

In this paper, we proposed the nonlinear cubic step function $p(s) = 4(s-1/2)^3 + 1/2$ instead of the linear step function $p(s) = s$ used in Das et al.[5] in the adiabatic quantum computation to effectively solve Deutsch problem. We could obtain the higher observation probability of finding the solution than that of Das et al.[5].

It seems that the most suitable step function will change depending on the number of qubit and the target problem like 3-SAT problem. Therefore, we must try to do the similar numerical experiment to check whether the cubic step function is appropriate or not.

Acknowledgements. This work was supported by Grant-in-Aid for Scientific Research (22500137) from the Ministry of Education, Culture, Sports, Science and Technology, Japan.

References

1. Farhi, E., Goldstone, J., Gutmann, S., Sipser, M.: Quantum Computation by Adiabatic Evolution, quant-ph/0001106 (2000)
2. Farhi, E.: A quantum adiabatic evolution algorithm applied to random instances of an NP-complete problem. Science 292, 472–476 (2001)

3. Van, W., Mosca, D.M., Vazirani, U.: How Powerful is Adiabatic Quantum Computation? quant-ph/0206003 (2002)
4. Hogg, T.: Adiabatic Quantum Computing for Random Satisfiability Problems, quant-ph/0206059 (2002)
5. Das, S., Kobes, R., Kunstter, G.: Adiabatic Quantum Computation and Deutsch's Algorithm, quant-ph/0111032 (2001)
6. Pati, A.K., Braunstein, S.L.: Deutsch-Jozsa algorithm for continuous variables, quant-ph/0207108v1 (2001)
7. Wei, Z., Ying, M.: A modified quantum adiabatic evolution for the Deutsch-Jozsa problem. Physics Letters A 354, 271–273 (2006)

A Computational Model of Imitation and Autonomous Behavior in Continuous Spaces

Tatsuya Sakato, Motoyuki Ozeki, and Natsuki Oka

Abstract. Learning is essential for an autonomous agent to adapt to an environment. One method that can be used is learning through trial and error. However, it is impractical because of the long learning time required when the agent learns in a complex environment. Therefore, some guidelines are necessary to expedite the learning process in the environment. Imitation can be used by agents as a guideline for learning. Sakato, Ozeki and Oka (2012) proposed a computational model of imitation and autonomous behavior. In the model, an agent can reduce its learning time through imitation. In this paper, we extend the model to continuous spaces, and add a function for selecting a target action for imitation from observed actions to the model. By these extension and adaptation, the model comes to adapt to more complex environment. Even in continuous spaces, the experimental results indicate that the model can adapt to an environment faster than a baseline model that learns only through trial and error.

Keywords: reinforcement learning, imitation, continuous spaces.

1 Introduction

Learning is essential for an autonomous agent to adapt to an environment. An agent can adapt to the environment by learning through trial and error. One way of learning through trial and error is by reinforcement learning[1]. However, adapting to a complex environment without any guidelines is a very time-consuming process[2]. Therefore, some guidelines are necessary for speeding-up learning in a complex environment. The use of appropriate

Tatsuya Sakato · Motoyuki Ozeki · Natsuki Oka
Dept. of Information Science,
Kyoto Institute of Technology, Kyoto, Japan
e-mail: sakato@ii.is.kit.ac.jp, {ozeki,nat}@kit.ac.jp

R. Lee (Ed.): *SNPD*, SCI 492, pp. 37–51.
DOI: 10.1007/978-3-319-00738-0_4

guidelines allows an agent to shorten its learning time[2][3][4]. Agents that have already adapted to an environment would perform actions appropriate to the environment. Therefore, imitation of the behavior of other agents that have already adapted to the environment would help an agent that has not adapted to the environment to shorten its learning time, and this imitation of behavior could act as a guideline for learning.

There are five key questions when an agent imitates behavior of other agents: "who to imitate", "when to imitate", "what to imitate", "how to imitate" and "how to evaluate a imitation"[5][6][7].

Sakato, Ozeki and Oka[8] focused on the two questions: "what to imitate" and "how to evaluate an imitation", and they proposed a computational model of imitation and autonomous behavior.

"What to imitate" was realized by selecting a policy for imitation. The way of imitation depended on policies for imitation when an agent tries to imitate an observed action. For example, they considered the case in which an agent tries to imitate the action of picking up an apple on a table by another agent. Then, policies the agent could adopt were "imitating movements of the hand", "picking up the same object", "picking up the same kind of object", and so on. The agent selected a policy from among these candidates.

"How to evaluate an imitation" was realized by evaluating an action the agent had performed when the agent had imitated an action of the other agent. For example, they considered the case in which the agent observed the other agent eating an apple the eating agent had, and tried to imitate the action with policy "eating an object which is held by itself". Then, the agent that had performed the action as imitation should have evaluated whether the policy the agent had selected was appropiate or not. In this case, policy "eating an object which is held by itself" was not appropiate if the agent had an inedible object. In this case, the agent should have selected policy "eating the same kind of object".

The agent trying to imitate should select a policy for imitation. In Sakato et al.'s model[8], policies were defined in advance, and each policy was evaluated by reward the agent had got. In their model, the purpose of the learning agent was to learn appropriate autonomous bahavior by imitation learning. Reinforcement learning and imitation learning were used for learning appropiate autonomous bahavior in their model. The learning agent performed actions which are determined from reinforced actions and observed actions, and learned appropriate autonomous bahavior.

In Sakato et al.'s model[8], discrete state and action spaces were used for learning. In order to apply for a real robot, their model should be extended to continuous state and action spaces for learning. Moreover, an observed action imitated was only the last observed action in their model, however, for the more appropriate imitation, targets to imitate should be a series of observed actions. Therefore, in this paper, we extend their model to continuous spaces, and add a function for selecting a target action for imitation from observed actions to their model.

The remainder of this paper is organized as follows. The related works are described in Section 2. The proposed computational model of imitation and autonomous behavior is described in Section 3. In Section 4, the evaluation experiments of the proposed model are described. In Section 5, the results of the experiments are described, and the discussion is provided. Finally, the conclusion is presented in Section 6.

2 Related Works

Ng et al. proposed a model that could shorten the learning time of the agent by using the shaping reward based on states[3]. Wiewiora et al. proposed a model using the shaping reward not based on states, but based on states and actions[2]. Tabuchi et al.[9] proposed a method of acquisition of behaviors by reinforcement learning using the shaping reward based on the result of imitation learning, applying the study by Ng et al.. The methods like these needed in advance to determine a policy for imitation, and to prepare the shaping rewards. The methods of selecting a policy for imitation were studied by Alissandrakis et al.[6], Billard et al.[5], and so on. Alissandrakis et al. proposed a method of selecting a policy from policies for imitation, and experimented the method in a chess world. In the chess world, an imitator agent tended to reproduce either subparts or the complete path followed by the demonstrator. Billard et al. applied Alissandrakis et al.'s method to learning by imitation of tasks involving object manipulation and gestures.

In the study by Tabuchi et al., the more frequent the state was observed, the more the shaping reward was given. In the study by Alissandrakis et al. or Billard et al., the more similar an performed action which an agent selected using a policy for imitation was to the observed action, the more frequently the agent selected the policy.

In contrast with these, in Sakato et al.'s model[8], the agent selected a policy for imitation depending on the reward the agent had got using the policy. In Sakato et al.'s model[8], the purpose of the learning agent was not reproduction of the observed movements, but learning appropriate autonomous behavior. Therefore, the learning agent should have selected a policy for imitation such that the learning agent could get more reward by imitation rather than a policy for imitation such that the learning agent could perform more similar action to the observed action. In Sakato et al.'s model[8], policies for imitation ware prepared in advance. In order to select a policy for imitation, a value was given to each policy. Each value reflects importance of each policy. When the learning agent observes actions the other agent performed, the learning agent selected a policy for imitation based on the values. In Sakato et al.'s model[8], the learning agent came to select a policy for imitation such that the learning agent could get more reward by updating values based on rewards the learning agent had got by imitation. The tendency to perform an observed action depended on similarity between the features of a state which ware defined in advance (hereafter, state features) of the observed agent and the present state features of the learning agent. A value was given to each feature, and the values influenced similarity between features.

In this paper, we extend the model to continuous spaces, and adapt the model to select an imitated target from observed actions. For these extension and adaptation, we make the following extension.

- Adapting Sakato et al.'s model[8] to continuous state features and continuous action features (which are features of an action)
- Adapting Sakato et al.'s model[8] to reinforcement learning in continuous state spaces and continuous action spaces
- Adapting the similarity to continuous state spaces
- Adapting Sakato et al.'s model[8] to select a target action for imitation from the observed actions
- Adding a function for selecting a target action for imitation from the observed actions to Sakato et al.'s model[8]
- Adapting Sakato et al.'s model[8] to update values of continuous state features and continuous action features

The specification of each extension is described in Section 3.

3 Proposed Model in Continuous Spaces

It assumes that there are an agent which performs optimal actions (hereafter, optimal agent) and a learning agent in the environment. The proposed model is implemented to the learning agent. When the optimal agent performs actions, the learning agent observes the actions and states in which the actions ware observed. The learning agent determines an action to perform from the observed actions and a reinforced action (see Fig. 1). In this paper, the proposed model is applied to learning in the environment that involoves continuous spaces. The specification of the proposed model is described below.

3.1 Learning Agent

In the proposed model, the learning agent consists of the reinforcement learning (RL) module, the imitation module, and the action selection module (Fig. 1). The agent learns autonomous behavior through trial and error by reinforcement learning in a given environment. Learning autonomous behavior through trial and error by reinforcement learning is carried out by the RL module. The agent tends to imitate actions performed by the optimal agent if there is the optimal agent in the environment. Observing actions performed by the optimal agent and selecting a policy for imitation, which is used to determine "what to imitate", are carried out by the imitation module. The action selection module determines whether the agent performs a learned action or imitates an action performed by the optimal agent.

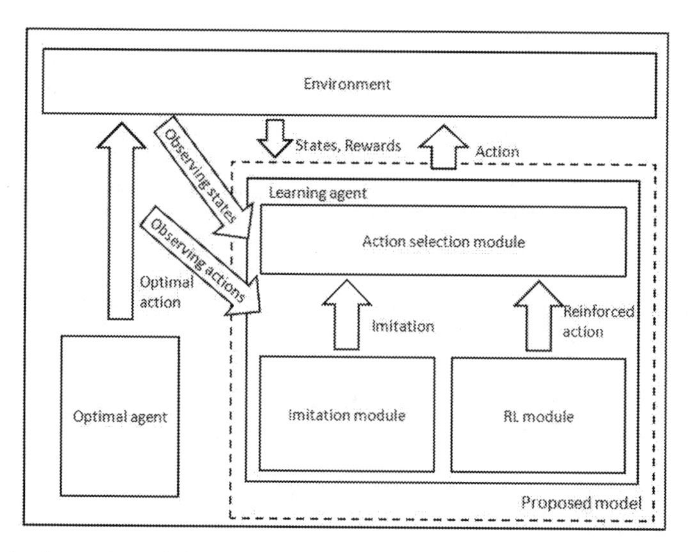

Fig. 1 Configuration of the proposed model

The learning and action generation process of the learning agent is described below.

1: **if** the optimal agent is performing an action, **then**
2: Observe a state and an action.
3: **else**
4: Select a target action from the observed actions.
5: Choose between two alternatives: a reinforced action or an action performed by the optimal agent with a possibility proportional to the similarity between a state in which the target action was observed and the present state of the learning agent.
6: **if** the learning agent chose an action performed by the optimal agent, **then**
7: Select a policy for imitation with a possibility proportional to the values of the policies.
8: Imitate the target with the selected policy.
9: **else**
10: Perform the reinforced action.
11: **end if**
12: Get rewards.
13: Perform reinforcement learning.
14: Update the values.
15: **end if**
16: Go to 1.

The specification of each process is described below.

3.2 Reinforcement Learning

Reinforcement learning is used in the RL module in order to learn autonomous behavior. Function approximators are used for approximations of values of states and actions. In the proposed model, we use CACLA by van Hasselt[10] as a reinfocement learning method for learning in continuous spaces.

3.3 Observing and Recording States and Actions

When the optimal agent performs an action, the learning agent recognizes an action with the action repertoire of the learning agent itself, and records the action with states in which the action was observed. The observed states and actions are recorded as a sequence of states and actions. States and actions are represented by a set of features such as "the type of an object", "the location of an object", and "whether the learning agent holding an object or not". A recorded set of state and action features in the sequence can be a target for imitation. States and actions are recorded by the imitation module.

3.4 Selecting a Target to Imitate

When the learning agent imitates the observed action, a target to imitate is an action such that a state in which the action was performed is the most similar to the present state of the learning agent. $\mathit{similarity}(s, s')$ is defined as the similarity between states as follows:

$$\mathit{similarity}(s, s') = \frac{\sum_{i=0}^{n_s} \exp(V_s(i)) f_i(s_i, s'_i)}{\sum_{i=0}^{n_s} \exp(V_s(i))} \tag{1}$$

where s and s' are states, s_i and s'_i are i-th state features of s and s', respectively, n_s is the number of the state features, and $V_s(i)$ is the value of i-th state feature. $V_s(i)$ is updated through learning. The updating process is described in Section 3.7. $f_i(s_i, s'_i)$ is a similarity between s_i and s'_i. The definition of the similarity between the state features is described below. A target to imitate is an action such that a state in which the action was observed is the most similar to the present state of the learning agent. Therefore, *target*, which is a target to imitate, is selected as follows:

$$\mathit{target} = \arg\max_j \mathit{similarity}(s, s^j) \tag{2}$$

where s^j is the j-th element of the sequence of the observed states.

There are continuous features and discrete features. $sim_c(x, y)$, similarity between continuous features x, y is defined as follows:

$$\mathrm{sim}_c(x, y) = \exp(\frac{-|x - y|^2}{2c^2}) \tag{3}$$

$sim_d(x, y)$, similarity between discrete features x, y is defined as follows:

$$\mathrm{sim}_d(x, y) = \delta_{xy} \tag{4}$$

where δ_{xy} is the Kronecker delta, which means

$$\delta_{xy} = \begin{cases} 1 \ (x = y) \\ 0 \ (x \neq y) \end{cases} \tag{5}$$

In Eq. (1), $f_i(s_i, s'_i)$, which is a similarity between s_i and s'_i, can be $\mathrm{sim}_c(x, y)$ or $\mathrm{sim}_d(x, y)$.

3.5 Selecting a Policy for Imitation

A policy for imitation, which is used to determine "what to imitate", is selected by the imitation module.

When the learning agent imitates the selected action, the learning agent selects a policy for imitating the selected action from n_a policies. The probability of selecting policy k is defined as follows:

$$\Pr(k) = \frac{\exp(V_a(k))}{\sum_{k=0}^{n_a} \exp(V_a(k))} \tag{6}$$

where $V_a(k)$ is the value of policy k, and is updated through learning.

3.6 Action Selection of the Learning Agent

An action performed by the learning agent is selected by the action selection module.

The learning agent determines whether it performs a reinforced action or imitates an action performed by the optimal agent. The determination is based on the similarity between a state in which the target action was observed and the present state of the learning agent.

When the learning agent selects the ith element of the sequence of the observed states and actions, the probability $\Pr(imitation)$ that the learning agent imitates the observed action is defined as follows:

$$\Pr(imitation) = similarity(s, s^{\mathrm{target}}) \tag{7}$$

where s is the present state of the learning agent, and s^{target} is the *target*-th element of the sequence of the observed states. The probability Pr(*autonomous*) with which the learning agent performs the reinforced action is defined as follows:

$$\Pr(autonomous) = 1 - similarity(s, s^{\text{target}}) \tag{8}$$

When the learning agent imitates the observed action, the learning agent determines an action to perform with the method described in Section 3.5.

3.7 Updating Feature and Policy Values

We explained in Section 3.3 that the observed states and actions were recorded as a sequence of states and actions, and that an element of the sequence was recorded as a set of features of states and actions. When the learning agent imitates an element of the sequence of the observed states and actions, how to determine a set to imitate and a policy for imitation was described in Section 3.4 and Section 3.5. In the proposed model, values, which reflect the importance of features or policies, are given to state features and policies for imitation. Features contributes to selecting a target for imitation and a policy for imitation. The degree of the contribution of each feature depends on each feature value. The values should be learned from the environment, because the important state feature or the policy to adopt for imitation is different if the environment has changed. In the proposed model, the values are updated based on the action the learning agent performed and the reward the learning agent got. When the learning agent performed action a and got reward r, the values are updated as follows:

$$V_s^{t+1}(i) \leftarrow f_i(s_i, s_i')\alpha(f_i(s_i, s_i')r - V_s^t(i)) \tag{9}$$

$$V_a^{t+1}(i) \leftarrow g_i(a_i, a_i')\alpha(g_i(a_i, a_i')r - V_a^t(i)) \tag{10}$$

where $V_s^{t+1}(i)$ and $V_a^{t+1}(i)$ are the updated feature value and the updated policy value, respectively. $V_s^t(i)$ and $V_a^t(i)$ are the previous feature value and the previous policy value, respectively. $\alpha(0 < \alpha \leq 1)$ is the learning rate. a_i is the i-th feature of the action the learning agent performed, and a_i is the i-th feature of the target action of imitation. $f_i(s_i, s_i')$ and $g_i(a_i, a_i')$ are the similarities between state features and action features, respectively.

Reward is referenced to update each value. The learning adavances depending on similarity between features.

State values and policy values are updated by the imitation module.

4 Experiments

A learning experiment using the proposed model and a learning expriment using only the RL module as a baseline of the comparative study are performed.

4.1 Experimental Environment

The experiments are performed in the following environment. The performance of the proposed model is evaluated using a dining table simulator, which is shown in Fig. 2. There are heads of two agents and their right hands, a table, and one object in the environment. The types of objects are an apple, a bunch of bananas, and a stone. One of the objects is randomly placed on the table in the initial state. The agent at the bottom is the optimal agent, and the one at the top is the learning agent. The agents act each episode alternately. Each episode ends if the object in the environment is eaten or thrown away. The learning agent observes actions and states when the optimal agent acts. The learning agent performs actions and learns the optimal actions when the learning agent acts. The action determination processes were described in Section 3.

The agents can pick an object on the table, place an object on the table, eat an edible object, and throw away an object.

The agent is given reward 10.0 when the agent eats an edible object (an apple and a bunch of bananas). The agent is given reward 10.0 when the agent throws away a stone. The agent is given reward 0.0 when the agent performs any other actions.

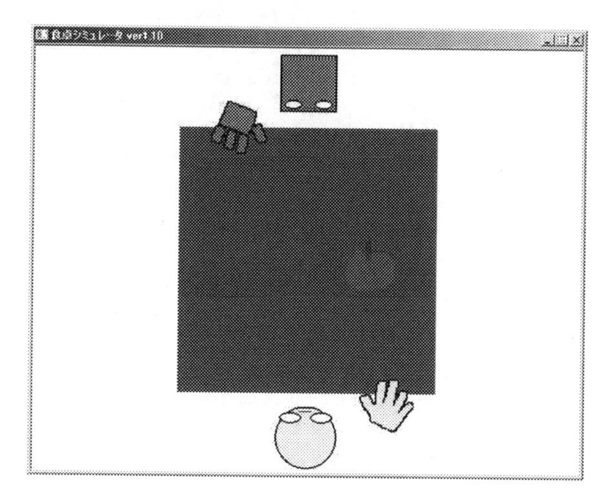

Fig. 2 Experimental environment

4.2 States

The states are expressed as a set of the following features. These features are used as parameters of function approximators for reinforcement learning.

- X-coordinate of the object
- Y-coordinate of the object
- Whether the object is an apple or not
- Whether the object is a bunch of bananas or not
- Whether the object is a stone or not
- Whether the agent is holding the object or not

"X-coordinate of the object" and "Y-coordinate of the object" are continuous features. The other features are discrete features (0 or 1).

4.3 Actions

An action of the learning agent is expressed as two-dimensinal coordinates. When the coordinates are indicated, the learning agent tries to reach out its hand to the coordinates. If the learning agent has nothing, the learning agent picks up an object at the coordinates. If the learning agent reached out its hand to the table, the learning agent picks up nothing. If the learning agent has something, and the coordinates of the action indicate somewhere on the table, the learning agent places the object at the coordinates. If the learning agent reached out its hand to the head of the learning agent, the learning agent tries to eat the object the learning agent is holding. If the learning agent reached out its hand to somewhere out of the table except the head of the learning agent, the learning agent throws away the object it has.

In the environment, the edible objects are an apple and a bunch of bananas, and the inedible object is a stone. If the learning agent tries to eat a stone, the learning agent fails to perform the action, and keep holding the stone.

4.4 Observed Features

When the optimal agent performs actions, the learning agent observes the following features of the optimal agent and the environment.

- State features
  - The type of an object in the environment, which can be an apple, a bunch of bananas, or a stone
  - Whether the optimal agent is holding an object or not
  - The relative location of the object
  - The absolute location of the object

- Action features
  - The place to where the optimal agent reached out its hand
  - The angle of shoulder of the optimal agent at the time the optimal agent performed the action
  - The type of an object at the place to where the optimal agent reached out its hand, which can be an apple, a bunch of bananas, a stone, a table, a head of the agent, or the outside of the table

The relative location is the location which uses the performing agent as the point of reference. The absolute location is the location which uses the environment as the point of reference. "The type of an object in the environment", "whether the optimal agent is holding an object or not" and "the type of an object at the place to where the optimal agent reached out its hand" are discrete features. "the absolute location", "the relative location", "the place to where the optimal agent reached out its hand" and "the angle of shoulder" are continuous features.

4.5 Policy for Imitation

The learning agent selects a policy from the following policies when the learning agent imitates an observed action.

- The same target
- The same angle
- The same place

If the learning agent selects policy "the same target", the learning agent reaches out its hand to an object which type is the same as that of an object the optimal agent reached out its hand to, e.g. when the learning agent observes that the optimal agent performs "picking an apple", then "eating", the learning agent performs actions to the coordinates of an apple, then to the coordinates of the head of the learning agent with policy "the same target". If the same target object is not on the table when the learning agent performs an action, policy "the same target" is not selected.

If the learning agent selects policy "the same angle", the learning agent performs an action with the same angle of shoulder of the learning agent as that of the optimal agent, e.g. when the learning agent observes that the optimal agent performs "picking an apple at the far right on the table from the optimal agent", the learning agent reaches out its hand to the far right on the table from the learning agent with policy "the same angle".

If the learning agent selects policy "the same place", the learning agent reaches out its hand to the place where the optimal agent reached out its hand, e.g. the learning agent observes that the optimal agent performs "reaching out its hand to $(0.5, 0.5)$, the learning agent reaches out its hand to $(0.5, 0.5)$ with policy "the same place". If the learning agent can not reach out its hand to the coordinates, policy "the same place" is not selected.

5 Results and Discussion

The learning agent performed 5000 steps in each experiment.

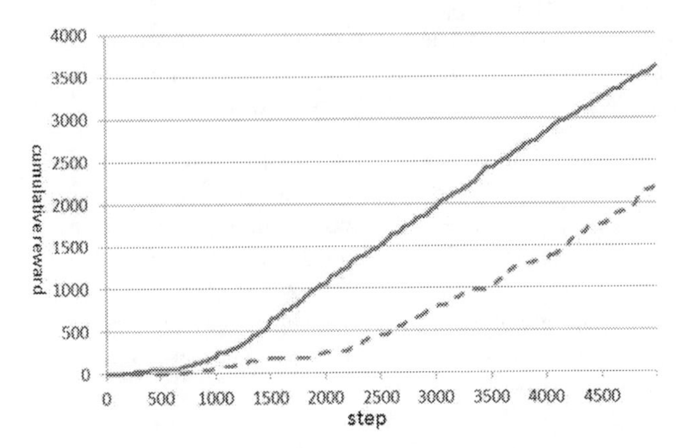

Fig. 3 Cumulative rewards observed in the experiments. The solid line represents the cumulative reward the learning agent got with the proposed model, and the broken line represents the cumulative reward the learning agent got only with the reinforcement learning method.

Fig. 3 shows the cumulative rewards observed in the experiments. The solid line represents the cumulative reward the learning agent got with the model proposed in this paper, and the broken line represents the cumulative reward the learning agent got only with the reinforcement learning method. The proposed model can adapt to the environment faster than a baseline model that only uses reinforcement learning. The results indicate that actions the optimal agent performed can be guidelines for the learning agent to which the proposed model is implemented.

Fig. 4 shows the change in values of state features, and Fig. 5 shows the change in the probability of selecting each policy.

Fig. 4 shows that the value of feature "the type of the object in the environment" is the highest among the four. The result indicates that the feature is important for the learning agent to learn in the environment. The learning agent tends to imitate the observed action when the learning agent observed the optimal agent performing an action to an object such that the type of the object is the same as an object in the environment.

The value of feature "whether the agent is holding the object or not" is the second highest among the four. The feature is also important for the learning agent to imitate action performed by the optimal agent, e.g. the agent cannot eat when the agent has nothing, or the agent cannot pick an object when the agent is holding something.

Feature "the absolute location of the object" and "the relative location of the object" are continuous features. The features are not as influential as

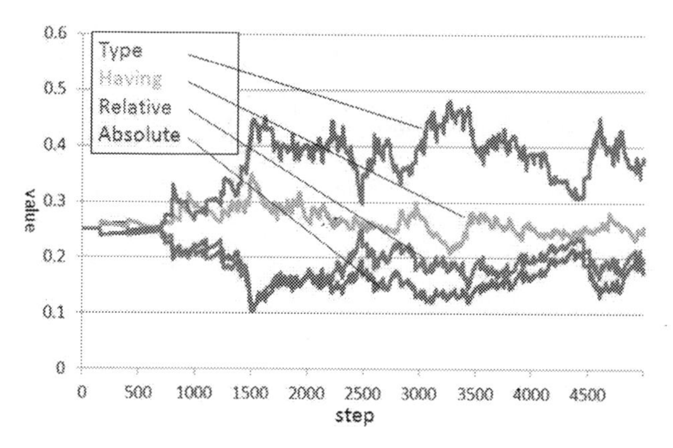

Fig. 4 Change in the values of the state features of the learning agent. "Type" means feature "the type of an object in the environment". "Having" means feature "whether the agent is holding the object or not". "Relative" means feature "the relative location of the object". "Absolute" means feature "the absolute location of the object".

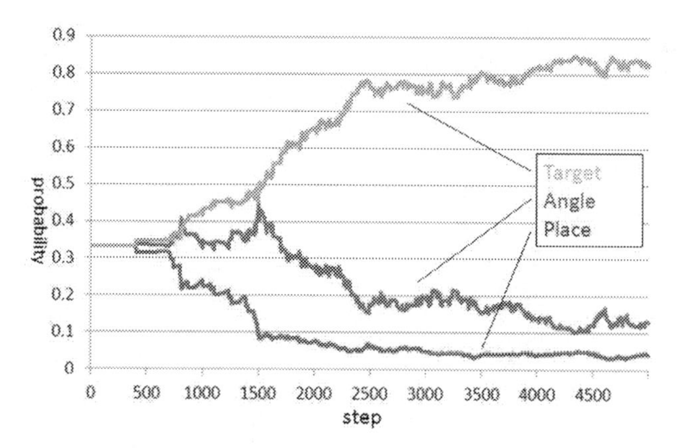

Fig. 5 Change in the probability of selecting each policy of the learning agent. "Target" means policy "the same target". "Angle" means policy "the same angle". "Place" means policy "the same place".

their values indicate because an object hardly appears at the same locations as once an object appeared at (see Eq. (1)).

In other words, in the case of the experiments, it seems that feature "the type of the object" and "whether the agent is holding the object or not" are the important features.

Fig. 5 shows that the value of policy "same target" is the very highest among the three. We believe this is because imitation with policy "the same target" is appropriate to the dining table environment. In addition, we believe that policy "the same angles" is a little higher than policy "the same place" because the appropriate action is an action such that the shoulder angle of the learning agent is the same as the observed angle when the learning agent performs the action after the learning agent picks an object (i.e. eating or thowing away).

The experimental results indicate that the learning agent can adapt to the dining table environment faster than a baseline model that learns only through reinforcement learning in the proposed model.

6 Conclusion

In this paper, Sakato et al.'s model[8] was extended to continuous state and action spaces. The model was implemented in the case of the dining table environment, then evaluated. The proposed model was evaluated based on the cumulative reward the learning agent got, changes in values of state features, and changes in probabilities for selecting policies for imitation. The learning agent that only uses reinforcement learning was also evaluated based on the cumulative reward the learning agent got as a baseline of the comparative study. The experimental results indicated that the proposed model can adapt to the environment faster than a baseline model that learns only through trial and error even in the environment with continuous state and action spaces. Moreover, values of the state features and probabilities of the policies for imitation also learned appropriately. In this paper, the learning agent that learns with the proposed model was only compared with the learning agent that learns only through reinforcement learning. However, we believe that the performance of the proposed model would change depending on whether learning of values of features is performed or not. Therefore, investigating the changes would be the aim of the next study.

The other challenge is to add a function for acquiring new behaviors to the learning agent. In the proposed model, the learning agent imitates the observed action with action repertoire of the learning agent itself. Therefore, the range of problems in which the learning agent can learn appropriate actions is fixed if action repertoire of the learning agent is designed in advance. In other words, complexity of problems in which the learning agent can learn appropriate actions depends on designs of state and action spaces, state and action features, and policies for imitation. A function to acquire new behaviors is necessary for more flexible imitation learning of the learning agent[11]. Moreover, workload during the design of the learning agent would be reduced if the learning agent can acquire new behaviors. In addition, we believe that the more appropriate behavior could be acquired if the learning agent acquires the concepts of actions from the environment rather than action repertoire of the learning agent is designed in advance.

References

1. Sutton, R., Barto, A.: Reinforcement Learning: An Introduction. MIT Press, Cambridge (1998)
2. Wiewiora, E., Cottrell, G., Elkan, C.: Principled methods for advising reinforcement learning agents. In: Proceedings of the Twentieth International Conference on Machine Learning (ICML 2003), pp. 792–799 (2003)
3. Ng, A., Harada, D., Russell, S.: Policy invariance under reward transformations: Theory and application to reward shaping. In: Proceedings of the Sixteenth International Conference on Machine Learning, pp. 278–287 (1999)
4. Price, B., Boutilier, C.: Implicit imitation in multiagent reinforcement learning. In: Proceedings of the Sixteenth International Conference on Machine Learning, pp. 325–334 (1999)
5. Billard, A., Epars, Y., Calinon, S., Schaal, S., Cheng, G.: Discovering optimal imitation strategies. Robotics and Autonomous Systems 47(2-3), 69–77 (2004)
6. Alissandrakis, A., Nehaniv, C., Dautenhahn, K.: Imitation with alice: Learning to imitate corresponding actions across dissimilar embodiments. IEEE Transactions on Systems, Man and Cybernetics, Part A: Systems and Humans 32(4), 482–496 (2002)
7. Nehaniv, C., Dautenhahn, K.: Of hummingbirds and helicopters: An algebraic framework for interdisciplinary studies of imitation and its applications. In: Interdisciplinary Approaches to Robot Learning, pp. 136–161 (1999)
8. Sakato, T., Ozeki, M., Oka, N.: A computatonal model of imitation and autonomous behavior. In: 2012 13th ACIS International Conference on Software Engineering, Artificial Intelligence, Networking and Parallel Distributed Computing (SNPD), pp. 13–18 (August 2012)
9. Tabuchi, K., Taniguchi, T., Sawaragi, T.: Efficient acquisition of behaviors by harmonizing reinforcement learning with imitation learning. In: The 20th Annual Conference of the Japanese Society for Artificial Intelligence (2006) (in Japanese)
10. van Hasselt, H.: Reinforcement learning in continuous state and action spaces. In: Reinforcement Learning: State of the Art, pp. 207–252. Springer (2012)
11. Kuniyoshi, Y.: Adaptive and emergent imitation as the fundamental of humanoid intelligence. Journal of the Robotics Society of Japan 25(5), 671–677 (2007) (in Japanese)

$DiaCTC(N)$: An Improved Contention-Tolerant Crossbar Switch

Jianfei Zhang, Zhiyi Fang, Guannan Qu*, Xiaohui Zhao, and S.Q. Zheng

Abstract. We recently proposed an innovative agile crossbar switch architecture called Contention-Tolerant Crossbar Switch, denoted as $CTC(N)$, where N is the number of input/output ports. $CTC(N)$ can tolerate output contentions instead of resolving them by complex hardware, which makes $CTC(N)$ simpler and more scalable than conventional crossbar switches. In this paper, we analyze the main factors that influence the performance of $CTC(N)$ and present an improved contention-tolerant switch architecture - Diagonalized Contention-Tolerant Crossbar Switch, denoted as $DiaCTC(N)$. $DiaCTC(N)$ maintains all good features of $CTC(N)$, including fully distributed cell scheduling and low complexity. Simulation results show that, without additional cost, the performance of $DiaCTC(N)$ is significantly better than $CTC(N)$.

1 Introduction

Crossbar is widely used in high-speed Internet switches and routers for its simplicity and non-blockingness. To simplify scheduling operations, variable size packets are segmented at input ports into fixed-size cells and reassembled at output ports.

Jianfei Zhang · Zhiyi Fang
College of Computer Science and Technology, Jilin University, China

Guannan Qu,
College of Computer Science and Technology, School of Communications Engineering, Jilin University, China
e-mail: qu.guannan@hotmail.com

Xiaohui Zhao,
School of Communications Engineering, Jilin University, China

S.Q. Zheng,
Department of Computer Science, University of Texas at Dallas, USA

* Corresponding author.

R. Lee (Ed.): *SNPD*, SCI 492, pp. 53–65.
DOI: 10.1007/978-3-319-00738-0_5

According to where packets (cells) are buffered, there are four basic types of crossbar switches, namely output queued (OQ), input queued (IQ), combined input and output queued (CIOQ), and crossbar with crosspoint buffered switches. In an OQ switch, cells arriving at input ports are forwarded to their destination output ports immediately and buffered in output queues. Without delay in input ports and switch fabric, OQ switches are powerful in terms of providing quality of services (QoS). Thus, theoretical studies on QoS guarantee are based on output queued switches[1]. Since an OQ switch requires memory speedup N, where N is the number of input/output ports of the switch, such QoS results are impractical.

The memory of IQ switches operates at the same speed as the external link rate and cells are queued in input ports. To avoid head-of-line (HOL) blocking problem, input buffers are arranged as virtual output queues (VOQs). Since it is hard to ensure QoS on IQ switches, CIOQ switches have been proposed as a trade-off design of OQ and IQ switches. In a CIOQ switch, the memory speed is S times faster than the link rate, where S is in the range of $1 < S < N$, and cells are buffered in both input ports and output ports. It was shown that a variety of quality of services are possible using CIOQ switches with a small constant S.

The performance of an IQ or CIOQ switch depends on scheduling algorithm, which selects contention-free cells and configures I/Q connections for switching cells in each time slot. For IQ switches, many scheduling algorithms based on maximum matching have been investigated (e.g. [2][3]). These scheduling algorithms provide optimal performance. Because the time complexity for finding maximum (size or weight) matchings is too high for practical use, heuristic algorithms for finding maximal matchings were considered instead (e.g. [4]-[8]). For a switch with N input ports and N output ports, such schedulers require $2N$ N-to-1 arbiters working in multiple Request-Grant-Accept (RGA) or Request-Grant (RG) iterations (which involve global information exchange) to obtain a maximal matching between inputs and outputs. Though implemented in hardware, these schedulers are considered too slow with too high cost for high-speed networks. The scheduling problem of CIOQ switches has also been considered. It was shown in [9] that, using an impractically complex scheduler, which implements the Stable Marriage Matching (SMM) algorithm [10], a CIOQ crossbar switch with a speedup of two in the switch fabric and memory can emulate an output queued (OQ) switch. This result is only theoretically important, because the SMM problem has time complexity $O(N^2)$.

To reduce scheduling complexity, crossbar switch with crosspoint buffers was proposed, which is also called buffered crossbar switch. Coordinating with input queues, crosspoint buffers decouple scheduling operations into two phases in each time slot. In the first phase, each input port selects a cell to place into a crosspoint buffer in its corresponding row, and in the second phase, each output port selects a crosspoint in its corresponding column to take a cell from. Input (resp. output) ports operate independently and in parallel in the first (resp. second) phase, eliminating a single centralized scheduler. Compared to unbuffered crossbars, the scheduling algorithms of buffered crossbars are much simpler. Considerable amount of work, e.g. [11]-[18], has been done on buffered crossbar with and without internal speedup.

However, N^2 crosspoint buffers take a large chip area, which severely restricts the scalability of buffered crossbar switches.

In summary, conventional crossbar switches, including crossbar with crosspoint buffers, require complex hardware to resolve output contentions. We recently proposed a new switch architecture called *contention-tolerant crossbar* switch, denoted by $CTC(N)$, where N is the number of input/output ports [19]. $CTC(N)$ tolerates output conflicts using a reconfigurable bus in each output column of the fabric. In this way, controllers distributed in input ports are able to operate independently and in parallel. This feature reduces the scheduling complexity and wire complexity, and makes $CTC(N)$ more scalable than conventional crossbar switches. $CTC(N)$ opens a new perspective on designing switches. This paper focuses on further discussion on $CTC(N)$, and presents an improved contention-tolerant switch architecture called *diagonalized contention-tolerant crossbar* switch, denoted as $DiaCTC(N)$. Simulation results show that, with *staggered polling* (SP) scheduling algorithms [20], $DiaCTC(N)$ significantly enhances the performance with the same low cost of $CTC(N)$.

2 Throughput Bottleneck of $CTC(N)$

In our previous work, we presented the $CTC(N)$ architecture. Similar to conventional crossbar, the fabric of $CTC(N)$ is comprised of N^2 crosspoints (Switching Element, SE) arranged as an $N \times N$ array. Each SE has three inputs, three outputs and two states, as shown in Figure 1 (a). Each input port i is equipped with a scheduler S_i. In one time slot, if input port i $(0 \leq i \leq N-1)$ wants to transmit a cell to an output port j $(0 \leq j \leq N-1)$, S_i sets the state of corresponding $SE_{i,j}$ to receive-and-transmit (RT) state. The remaining SEs in the same row be kept in cross (CR) state. If more than one input ports set their SEs as RT in the same output line (column), the output line is configured as a pipeline, as shown in Figure 1 (b). Cells transmitted from upstream input ports will be intercepted and buffered in downstream input ports. In this way, output contentions are tolerated in $CTC(N)$. Buffer in each input ports can be arranged as single FIFO queue or Virtual Output Queues according to queueing management policies, which contains cells both from outside of switch and from upstream input ports (if exist).

In [19], we theoretically proved that the throughput of $CTC(N)$ with single FIFO in each input ports and without speedup is bounded by 63%. To improve the performance of $CTC(N)$, we proposed *staggered polling scheduling algorithm scheme* (SP for short)[20]. In order to ease scheduling operations, buffer in input port i is arranged as N VOQs denoted by $VOQ_{i,j}$. S_i, the scheduler in input port i, maintains two sub-schedulers, i.e. the *primary sub-scheduler* PS_i and the *secondary sub-scheduler* SS_i, as shown in Figure 2. PS_i generates a unique number $c_i(t), 0 \leq c_i(t) \leq N-1$, in time slot t. That is, $c_{i'}(t) \neq c_{i''}(t)$ if $i' \neq i''$. At time t, if $VOQ_{i,c_i(t)}$ is not empty, the cell selected by PS_i is the HOL cell of $VOQ_{i,c_i(t)}$, denoted by $PS_i(t) = VOQ_{i,c_i(t)}$. Otherwise, PS_i returns null, denoted by $PS_i(t) = null$. SS_i maintains a set $L_i(t)$ of non-empty VOQ indices

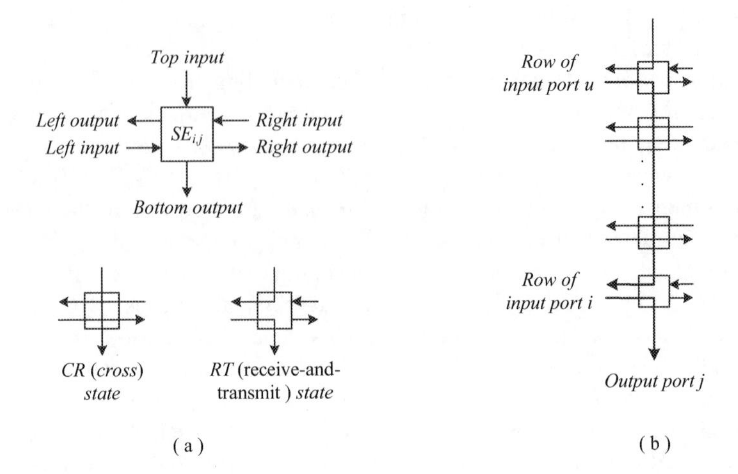

Fig. 1 (a) A crosspoint SE and its two states; (b) Each output line of $CTC(N)$ is a reconfigurable bus

(i.e. $L_i(t) = \{k|VOQ_{i,k}$ is not empty in time slot $t\}$), which is updated in every time slot. SS_i chooses one from $L_i(t)$ according to some algorithms as its scheduling result. Random pattern is one of representative secondary scheduling algorithm, i.e. SS_i picks one index of VOQ from $L_i(t)$ randomly. We use $q_i(t)$ to denote the index number of the VOQ chosen by SS_i in time slot t, i.e. $SS_i(t) = VOQ_{i,q_i(t)}$. If the $L_i(t)$ is empty, $SS_i(t) = null$.

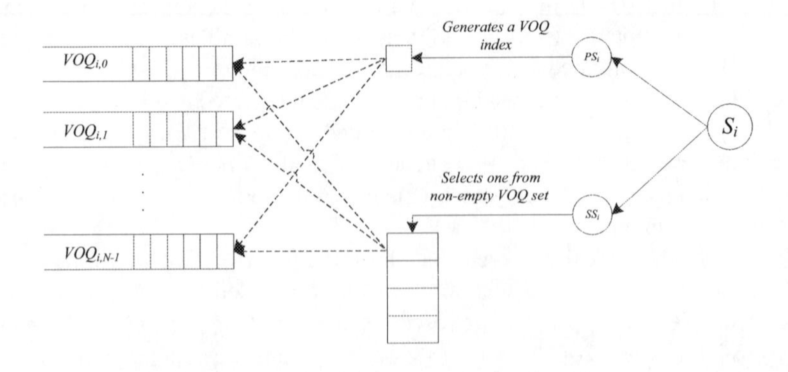

Fig. 2 Scheduling process in input port i

The two sub-schedulers operate in parallel. S_i chooses the scheduling result of PS_i first. If $PS_i(t) \neq null$, $S_i(t) = VOQ_{i,c_i(t)}$, which means $VOQ_{i,c_i(t)}$ will be serviced at time slot t. If $PS_i(t) = null$, S_i turns to consider the scheduler result of SS_i. If both PS_i and SS_i return null, $S_i(t) = null$, which means no VOQ will be serviced at time slot t.

In [20], we evaluated the performance of SP with several example scheduling algorithms. Simulation results showed that with zero knowledge of other input ports, the fully distributed schedulers "smartly" cooperated with each other and attained high performance. However, due to the intrinsic feature of $CTC(N)$, it is hard for the throughput of $CTC(N)$ to achieve 100%.

In what follows we analyze the main factors that affect the performance of $CTC(N)$. For easy analysis, let us consider the case with heavy offered load. Since each input port transmits one cells during one time slots, input port i can be modeled as a queue, denoted as Q_i, which contains cells from both outside and upstream and with N output destinations. $CTC(N)$ can be modeled as a queueing network as shown in Figure 3, where $a_i^o = \lambda$ is the arrival rate of Q_i from outside, and a_i^u is the arrival rate of Q_i from upstream input ports. r_i is the service rate of Q_i. For switch without internal speedup and under heavy traffic, we have $r_i = 1$.

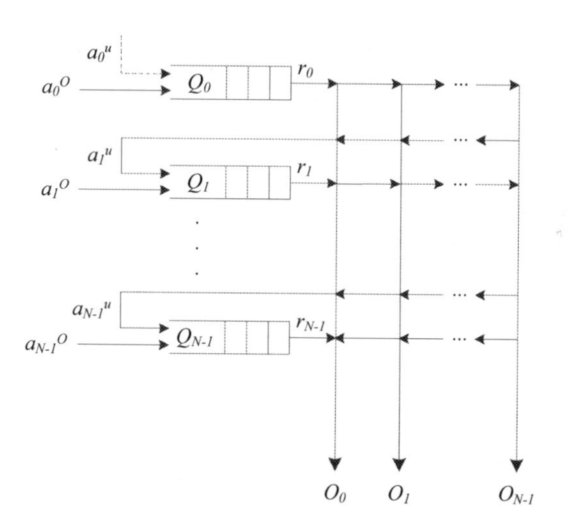

Fig. 3 Queueing model of $CTC(N)$ with single FIFO queue in each input port

For Q_i, the aggregate arrival rate a_i is:

$$a_i = a_i^o + a_i^u = \lambda + a_i^u. \tag{1}$$

When $a_i > r_i$, Q_i is overloaded and the throughput of $CTC(N)$ fails to achieve 100%.

Let $a_{i,j}^u$ be the arrival rate of Q_i from upstream input ports with output port j as its destination. We have

$$a_i^u = \begin{cases} 0 & \text{if } i = 0; \\ \sum_{j=0}^{N-1} a_{i,j}^u & \text{if } 0 < i \leq N-1. \end{cases} \tag{2}$$

From the analysis in [21], we know that

$$a^u_{i,j} > a^u_{k,j} \tag{3}$$

for $i > k$ and $0 \leq j \leq N-1$. It implies,

$$a^u_i > a^u_k \tag{4}$$

for $i > k$. Clearly, the aggregate upstream traffic increases as input port i increments. In the worst case with $\lambda = 1$, from Equation (1) and (2), we have:

$$a_i = \begin{cases} 1 & \text{if } i = 0; \\ 1 + \sum_{j=0}^{N-1} a^u_{i,j} & \text{if } 0 < i \leq N-1. \end{cases}$$

Obviously, even Q_1 is overloaded for $a_1 > r_1$. With the same service rate and multiple times heavier arrivals, downstream input ports suffer from more severe overload.

SP scheduling algorithm scheme was designed for reducing upstream arrivals by diminishing interceptions. With staggered polling pattern, the primary sub-schedulers select cells to form a conflict-free I/O matching. In order to maximize the utilization of input ports, the second sub-schedulers select cells arbitrarily, which may cause conflicts and interceptions. Simulation results in [20] showed that SP algorithms successfully enhance the throughput. However, the structure of $CTC(N)$ dictates that Equations (3) and (4) still hold with SP algorithms. Unbalanced upstream arrivals lead to overloading traffic for downstream input ports. It explains the phenomenons in [20] that the throughput began to go down when offered load $\lambda = 0.5$, where the input port $N-1$ who had heaviest upstream traffic started to be overloaded.

3 Diagonalized Contention-Tolerant Crossbar Switch Architecture

In order to improve the throughput of $CTC(N)$, we introduce an improved $CTC(N)$ architecture called *Diagonalized Contention-Tolerant Crossbar Switch*, denoted as $DiaCTC(N)$. $DiaCTC(N)$ is exactly the same in all aspects of $CTC(N)$, except the connections in each SE column.

In $CTC(N)$, SEs in each output column form a unidirectional alignment. $SE_{i',j}$ is an upstream node of $SE_{i'',j}$ in output column j, where $0 \leq i' < i'' \leq N-1$ and $0 \leq j \leq N-1$. Therefore, input port 0 is the top input for any output destination and it only has traffic from outside. A cell transmitted out from input port 0 could be intercepted by $N-1$ downstream input ports. Input port $N-1$ is the bottom input for any output destination. Cells buffered in input port $N-1$ are from outside and $N-1$ possible upstream input ports by N possible output columns. While in $DiaCTC(N)$, consider output column j, $0 \leq j \leq N-1$. SEs in output column j are classified into the following three classes:

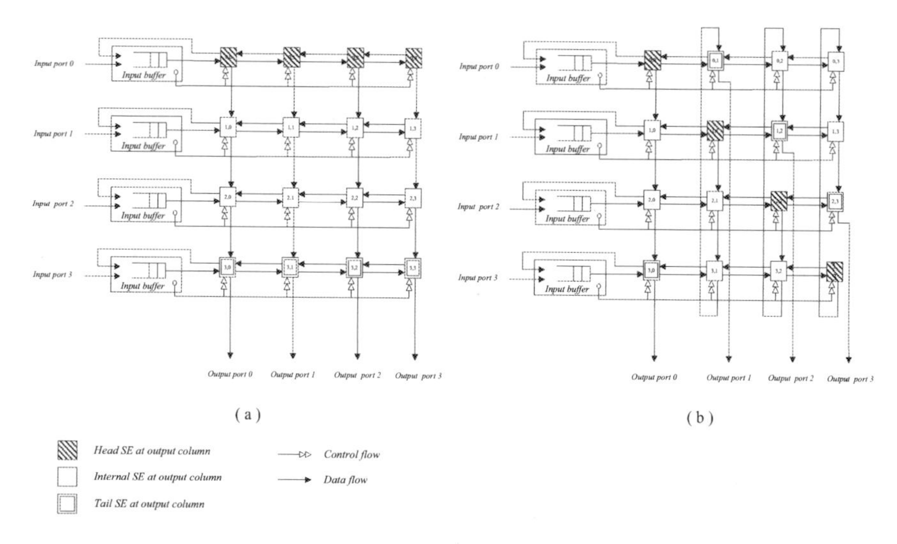

Fig. 4 (a) $CTC(4)$ architecture; (b) $DiaCTC(4)$ architecture

- *Head SE:* $SE_{i,j}$ is the head SE of output column j when $i = j$. The associated I_i is the top input for output destination O_j. Cell transmitted from I_i to O_j possibly is intercepted by I_d, $d = i + k \bmod N$ and $1 \leq k \leq N - 1$.
- *Tail SE:* $SE_{i,j}$ is the tail SE when $i = j - 1 \bmod N$. The associated I_i is the bottom input for output destination O_j. Cells transmitted from I_i to O_j arrives at O_j without being intercepted.
- *Internal SE:* $SE_{i,j}$ is an Internal SE when $i = j + k \bmod N$, $1 \leq k \leq N - 2$. I_i has upstream inputs and downstream inputs for output destination O_j. A cell transmitted from I_i to O_j could be intercepted by its downstream input I_d, i.e. $d = i + k \bmod N$ and $1 \leq k \leq (j - 1 - i) \bmod N$.

Let C_i be the aggregation of cells which might be buffered in I_i. $c_{s,d}$ is the cell which originally arrived at I_s from outside with O_d as its destination. $c_{s,d} \in C_i$ and it satisfies following condition:

$$s = \begin{cases} i & \text{if } d = i; \\ (i - k) \bmod N, 0 \leq k \leq (i - d) \bmod N & \text{otherwise.} \end{cases}$$

Figure 4 (b) shows a $DiaCTC(4)$, and its counterpart $CTC(4)$ is shown in Figure 4 (a) with SEs serving as Heads and Tails being labeled. In $CTC(N)$, all SEs in the top row are Head SEs and SEs in the bottom row are Tail SEs, while in $DiaCTC(N)$ there is exactly one Head SE and exactly one Tail SE in each row.

Compared with $CTC(N)$, $DiaCTC(N)$ balances the aggregate upstream traffics over all input ports without additional hardware cost, and Equations (3) and (4) don't hold. In the next section, we show that better performance will be achieved by this simple, but meaningful, modification of $CTC(N)$.

4 Performance Evaluation

The performance of *DiaCTC*(32) and *CTC*(32) with Staggered Polling (SP) algorithm scheme are compared in terms of mean cell delay under uniform traffic and non-uniform traffic by simulations. Random pattern is chosen as an example secondary scheduling algorithm in SP algorithm scheme. We also consider the well-known iSLIP method of one iteration for 32×32 conventional crossbar switch for the reason that only one iteration may be performed in each time slot in cell switching at the line speed.

4.1 Uniform Traffic

For uniform traffic, the traffic distributed over all output destinations uniformly. We test the performance use Bernoulli and Bursty arriving pattern. With Bernoulli arrivals, *DiaCTC*(32)-SP-Random performs the same performance with iSLP when offered load $\lambda \leq 0.55$, and has smaller mean cell delay than iSLIP when $0.55 \leq \lambda \leq 1$, as shown in Figure 5. Obviously, compared to the performance of *CTC*(32)-SP-Random, *DiaCTC*(32)-SP-Random has remarkable improvement.

Figure 6 illustrates the performance under bursty arrivals with burst length are 16, 32 and 64. The performance of *DiaCTC*(N)-SP-Random, *CTC*(N)-SP-Random and iSLIP decrease slightly with increasing burst length at the same offered load λ, and the performance decline of iSLIP is more evident than the other two. It implies that *CTC*(N) and *DiaCTC*(N) switches are affected less than iSLIP by

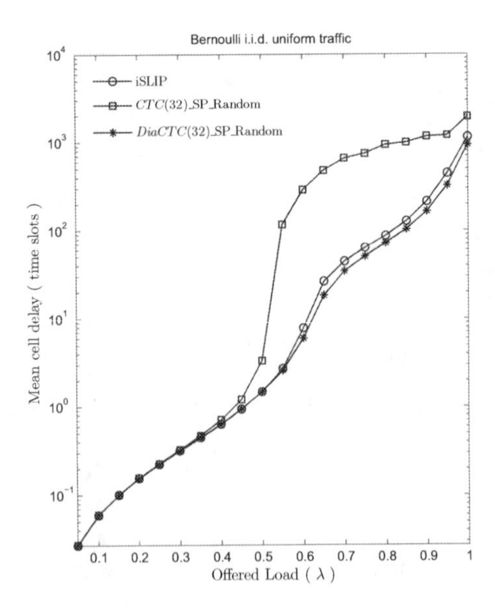

Fig. 5 Mean cell delay under Bernoulli i.i.d. uniform traffic

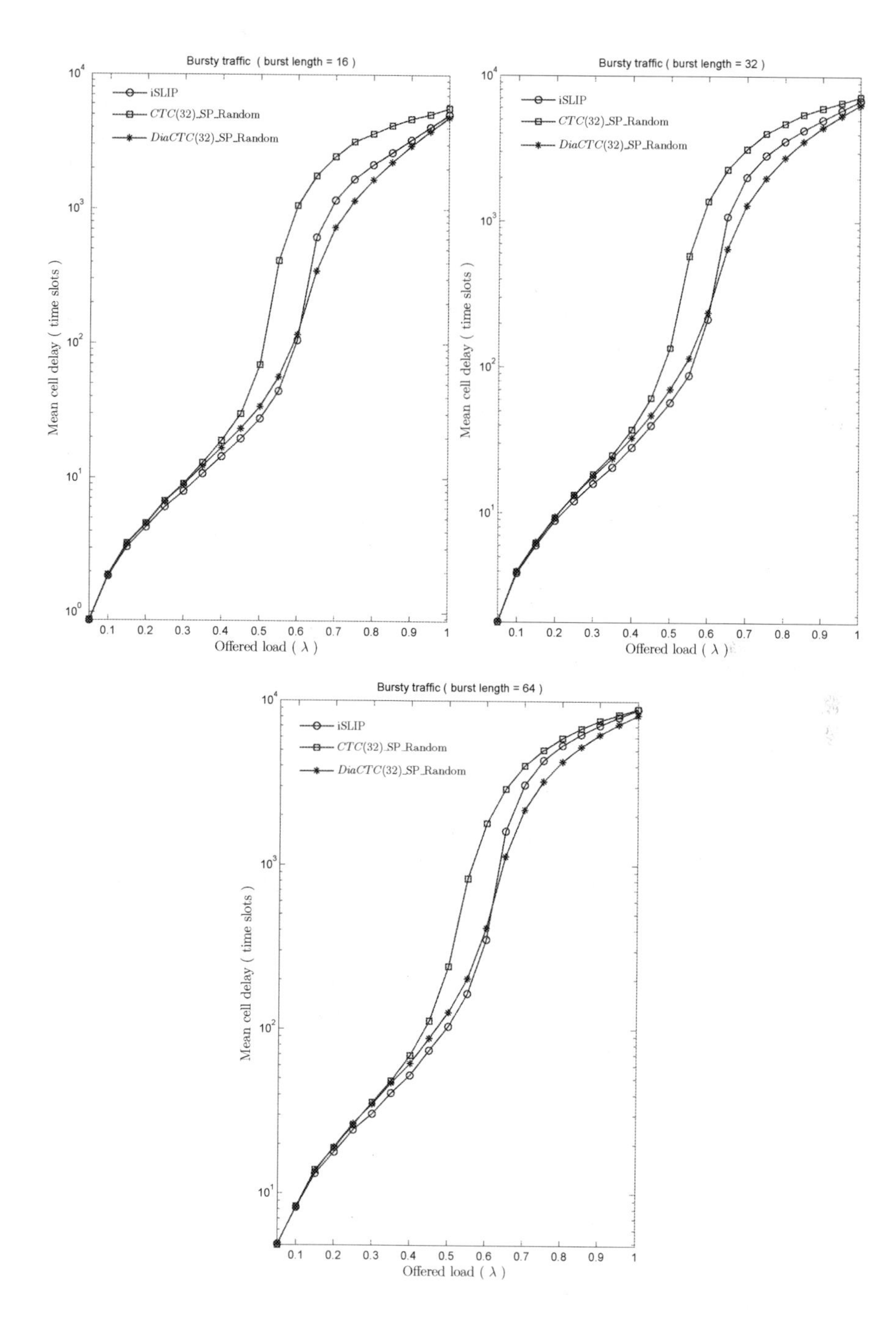

Fig. 6 Mean cell delay under bursty traffic with different burst length

burst length. The iSLIP outperforms *DiaCTC*(32)-SP-Random and *CTC*(32)-SP-Random in these three graphs when $\lambda \leq 0.6$, however, *DiaCTC*(32)-SP-Random shows the best performance when $0.6 < \lambda \leq 1$. *CTC*(32)-SP-Random has the similar performance with *DiaCTC*(32)-SP-Random and iSLIP at $\lambda = 1$ with minor difference.

4.2 Non-uniform Traffic

We chose three schemes from several nonuniform traffic models: Asymmetric [22], Chang's [23], and Diagonal [24]. Let $\lambda_{i,j}$ be the offered load arriving at input port i and forwarding to output port j. Asymmetric traffic model is defined as

$$\lambda_{i,(i+j)\text{mod}N} = \lambda a_j,$$

where $a_0 = 0$, $a_1 = (r-1)/(r^N - 1)$, $a_j = a_1 r^{j-1} \forall j \neq 0$, and $\lambda_{i,j}/\lambda_{(i+1)\text{mod}N,j} = r, \forall i \neq j, (i+1)\text{mod}N \neq j$, $r = \lambda_{\text{min}}/\lambda_{\text{max}} = a_{N-1}/a_1 = r^{-1/(N-2)}$. Chang's traffic model is defined as

$$\lambda_{i,j} = \begin{cases} 0 & \text{if } i = j; \\ \frac{\lambda}{N-1} & \text{otherwise.} \end{cases}$$

Diagonal traffic model has very skewed loading which defined as

$$\lambda_{i,j} = \begin{cases} \frac{2\lambda}{3} & \text{if } i = j; \\ \frac{\lambda}{3} & \text{if } j = i+1; \\ 0 & \text{otherwise.} \end{cases}$$

The performance of *DiaCTC*(32) and *CTC*(32) with SP scheduling algorithm and iSLIP are given in figure 7. *DiaCTC*(32)-SP-Random shows the best performance with both asymmetric and Chang's arrivals. *CTC*(32) performs similar with *DiaCTC*(32) and is better than iSLIP when $0.8 \leq \lambda \leq 1$ under asymmetric traffic.

Under diagonal traffic, the mean cell delay of *CTC*(32)-SP-Random increases sharply when $0.65 \leq \lambda \leq 0.75$, and goes up slightly with $0.75 \leq \lambda \leq 0.8$. The delay of *DiaCTC*(32)-SP-Random and iSLIP rise smoothly with increasing offered load, and iSLIP outperforms *DiaCTC*(32)-SP-Random. *DiaCTC*(N) tends to scatter the traffic over all of the VOQs in one input by intercepting cells from other inputs. Thus SP scheduling algorithm operates well and the switch achieves high performance. It is good for the situation with heavy and more balanced traffic load, such as Bernoulli traffic, Bursty traffic, Asymmetric non-uniform traffic and so on. However, for diagonal traffic which only has cells forwarding to two output destinations in each input ports, load balancing process in *DiaCTC*(N) leads to unexpected delay. Even though *DiaCTC*(32)-SP-Random has slightly higher delay than iSLIP, considering its low arbitration complexity and fully distributed control feature, the performance is really prominent.

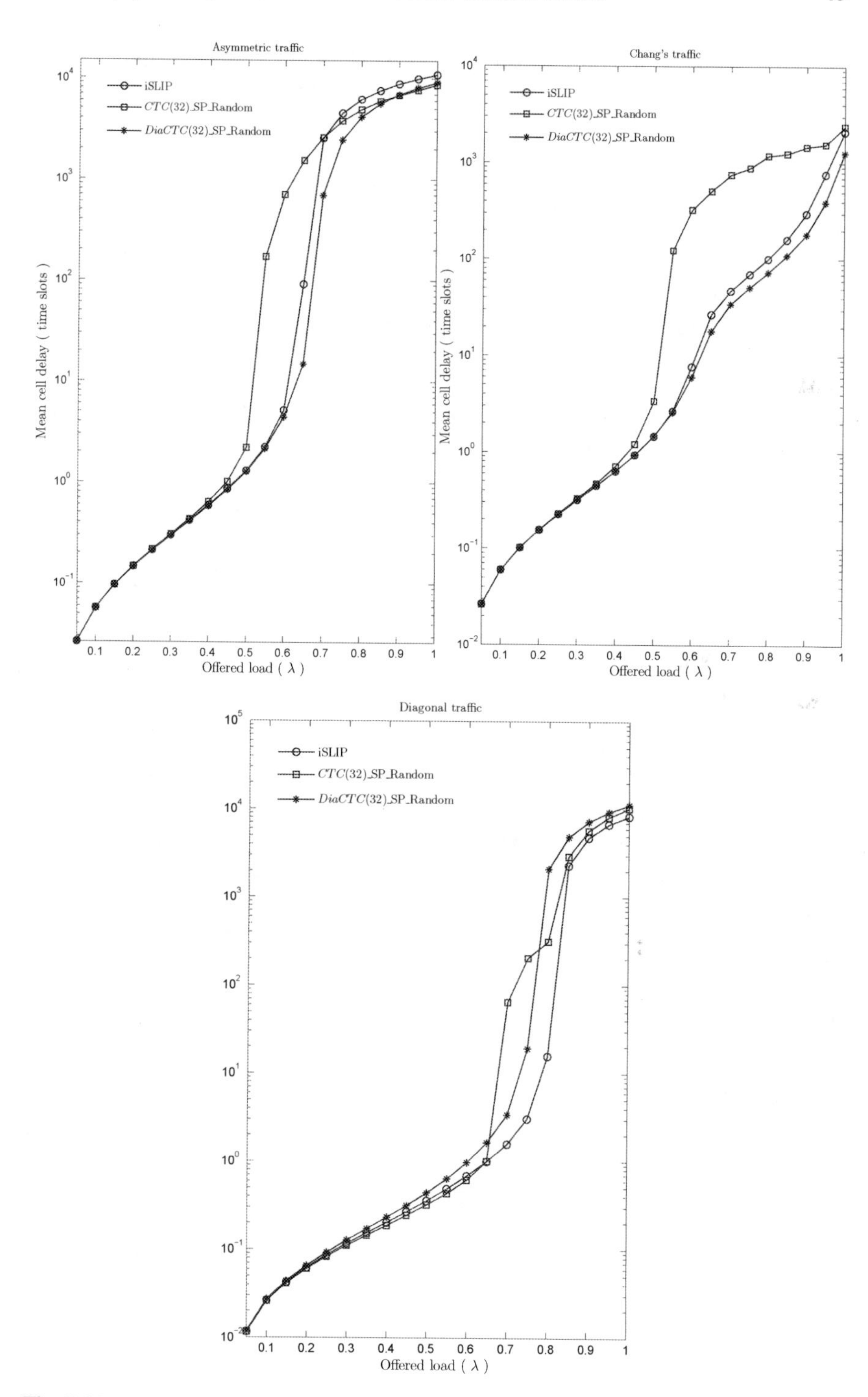

Fig. 7 Mean cell delay under three nonuniform traffic schemes

From above simulation results, we can conclude that $DiaCTC(32)$ significantly enhances the performance, but has the same good feature and low complexity as $CTC(N)$.

5 Concluding Remarks

In our previous work, we proposed an innovative agile crossbar switch architecture $CTC(N)$ and proved that its throughput of simple FIFO scheduling under Bernoulli i.i.d. uniform traffic is bounded by 63%. To improve performance, we proposed a fully distributed scheduling algorithm scheme called *staggered polling* (SP in short). Simulation results showed that, using SP scheduling algorithm scheme, the fully distributed schedulers "smartly" cooperate with each other and achieve high performance even with zero knowledge of other input ports.

This paper analyzes the main factor influencing performance of $CTC(N)$. We present an improved contention-tolerant crossbar switch called *diagonalized contention-tolerant crossbar*, denoted as $DiaCTC(N)$. Since $DiaCTC(N)$ has the same fully distributed control property and low complexity of $CTC(N)$, SP scheduling algorithms are able to operate on $DiaCTC(N)$ without any change. $DiaCTC(N)$ enhances the performance by balancing upstream traffic load for input ports. Simulation results show the outstanding improvement of performance of $DiaCTC(N)$ with SP scheduling algorithms.

$DiaCTC(N)$ illustrates a new approach to improving $CTC(N)$. However, out of sequence problem, which exists in $CTC(N)$, remains a challenging open problem for $DiaCTC(N)$. It can be reduced by designing sophisticated scheduling algorithms and queueing management methods. We remain the discussion of this problem in our subsequence papers. On the other hand, more algorithms can be designed for achieving good performance according to other QoS measures.

References

1. Zhang, H.: Service disciplines for guaranteed performance service in packet-switching networks. Proc. of IEEE 83, 1373–1396 (1995)
2. Hopcroft, J.E., Karp, R.M.: An $n^{5/2}$ algorithm for maximum matchings in bipartite graphs. SIAM, J. Comput. 2(4), 225–231 (1973)
3. McKeown, N., Mekkittikul, A., Anantharam, V., Walrand, J.: Achieving 100% throughput in an input-queued switch. IEEE Trans. on Commun. 47(8) (August 1999)
4. Tamir, Y., Chi, H.C.: Symmetric Crossbar arbiters for VLSI communication switches. IEEE Trans. on Parallel and Distributed Systems 4(1), 13–27 (1993)
5. Anderson, T.E., Owicki, S.S., Saxe, J.B., Thacker, C.P.: High speed switch scheduling for local area networks. ACM Trans. on Computer Systems 11(4), 319–352 (1993)
6. McKeown, N., Varaiya, P., Warland, J.: Scheduling cells in an input-queued switch. IEE Electronics Letters 29(25), 2174–2175 (1993)
7. McKeown, N.: The iSLIP scheduling algorithm for input-queued switches. IEEE/ACM Trans. on Networking 7(2), 188–201 (1999)
8. Chao, H.J.: Saturn: A terabit packet switch using dual round-robin. IEEE Commun. Magazine 8(12), 78–84 (2000)

9. Chuang, S.-T., Goel, A., McKeown, N., Prabhakar, B.: Matching output queuing with a combined input output queued switch. IEEE J. Select. Areas Commun. 17(6), 1030–1039 (1999)
10. Gale, D., Shapley, L.S.: Colleage admissions and the stability of marriage. Amer. Math. Monthly 69, 9–15 (1962)
11. Nabeshima, M.: Performance Evaluation of a Combined Input- and Crosspoint -Queued Switch. IEICE Trans. on Commun. E83-B(3) (March 2000)
12. Rojas-Cessa, R., Oki, E., Chao, H.J.: CIXOB-k: Combined input- and crosspoint-queued switch. IEICE Trans. on Commun. E83-B(3), 737–741 (2000)
13. Mhamdi, L., Hamdi, M.: MCBF: A high-performance scheduling algorithm for buffered crossbar switches. IEEE Commun. Letters (2003)
14. Chrysos, N., Katevenis, M.: Weighted fairness in buffered crossbar scheduling. In: Proc. of IEEE Workshop on High Performance Switching and Routing, pp. 17–22 (June 2003)
15. Rojas-Cessa, R., Oki, E., Chao, H.J.: On the combined input-crosspoint buffered switch with round-robin arbitration. IEEE Trans. on Commun. 53(11), 1945–1951 (2005)
16. Chuang, S.-T., Iyer, S., McKeown, N.: Practical algorithms for performance guarantees in buffered crossbars. In: Proc. of IEEE INFOCOM 2005, March 13-17, vol. 2, pp. 981–991 (2005)
17. Lin, M., McKeown, N.: The throughput of a buffered crossbar switch. IEEE Commun. Letters 9(5), 465–467 (2005)
18. He, S.-M., Sun, S.-T., Guan, H.-T., Zhang, Q., Zhao, Y.-J., Gao, W.: On guaranteed smooth switching for buffered crossbar switches. IEEE/ACM Trans. on Networking 16(3), 718–731 (2008)
19. Qu, G., Chang, H.J., Wang, J., Fang, Z., Zheng, S.Q.: Contention-tolerant crossbar packet switches. International Journal of Communication Systems 24, 168–184 (2011)
20. Qu, G., Chang, H.J., Wang, J., Fang, Z., Zheng, S.Q.: Designing fully distributed scheduling algorithms for contention-tolerant crossbar switches. In: Proc. of 11th International Conference on High Performance Switching and Routing, June 13-17 (2010)
21. Qu, G., Chang, H.J., Wang, J., Fang, Z., Zheng, S.Q.: Queueing analysis of multi-layer contention-tolerant crossbar switch. IEEE Commun. Letters 14(10), 972–974 (2010)
22. Schoene, R., Post, G., Sander, G.: Weighted arbitration algorithms with priorities for input-Queued switches with 100% throughput. In: Broadband Switches Symposium 1999 (1999)
23. Chang, C.-S., Lee, D.-S., Jou, Y.-S.: Load balanced birkhoff-von neumann switches. In: IEEE HPSR 2001, pp. 276–280 (April 2001)
24. Giaccone, P., Shah, D., Prabhakar, B.: An implementable parallel scheduler for input-queued switches. IEEE Micro. 22, 19–25 (2002)

Exegetical Science for the Interpretation of the Bible: Algorithms and Software for Quantitative Analysis of Christian Documents

Hajime Murai

Abstract. Systematic thought (such as Christian theology) has primarily been investigated using literature-based approaches, with texts that are usually more abstract and subjective in nature than scientific papers. However, as systematic ideas and thought influence all areas of human activity and thinking, the application of scientific methodologies such as bibliometrics, natural language processing, and other information technologies may provide a more objective understanding of systematic thought. This paper introduces four methods of quantitative analysis for the interpretation of the Bible in a scientific manner. The methods are citation analysis for interpreters' texts, vocabulary analysis for translations, variant text analysis for canonical texts, and an evaluation method for rhetorical structure. Furthermore, these algorithms are implemented for Java-based software.

Keywords: Bible, theology, interpretation, bibliometrics, NLP.

1 Introduction

As an aspect of higher cognitive functions, systematic thought has primarily been investigated using literature-based approaches, with texts that are usually more abstract and subjective in nature than scientific papers. However, as systematic ideas and thought influence all areas of human activity and thinking, the application of scientific methodologies such as bibliometrics, natural language processing, and other information technologies may provide a more objective understanding of systematic thought. By utilizing these new scientific methods,

Hajime Murai
Graduate School of Decision Science,
Tokyo Institute of Technology, Tokyo, Japan
e-mail: h_murai@valdes.titech.ac.jp

R. Lee (Ed.): *SNPD*, SCI 492, pp. 67–86.
DOI: 10.1007/978-3-319-00738-0_6

we can (a) ensure the objectivity and replication of results; (b) handle large-scale data precisely in a uniform manner [1].

I believe that it is possible to analyze the abstract thoughts and value systems embodied within a text corpus with such methods. In this paper, I focus on a Christian text corpus. Throughout history, traditional religions have exerted great influence on humanity. Most religions have certain canonical texts at their core, with the hermeneutics, or interpretations, of the canon also usually in text format. Thus, it is possible to represent key conceptualizations through the objective analysis of the canonical texts.

2 Approaches to Scientific Interpretation

As mentioned above, for the scientific analysis of thoughts, it is necessary that interpretations of the canon of target thoughts be analyzed scientifically. Unfortunately, it is currently impossible to achieve a scientific interpretation comparable to human interpretation, but it is possible to partially reproduce several human techniques of interpretation by utilizing scientific methods.

There are two quantitative approaches to interpreting the canon. The first is not a semantic interpretation of the canon itself, but an indirect approach to more clearly extract the details of interpretation. Although the second approach is direct, there are concerns that the resulting analysis is shallow because the canon itself does not always include as much information as is required in order to analyze the interpretation.

In the first, indirect approach, the relationship between the target text (the canon) and the texts that describe the interpretation of the target text (theologians' texts) is important. These relationships are called intertextuality. Citation analysis is an effective method for analyzing the relationships between texts. It clarifies the interpretation of some parts of the canon and the relationships between several parts of the texts. Therefore, this analysis is able to visualize the structure of interpreters' concepts. Citation analysis enables the scientific analysis of theological differences between theologians and between eras or sects [2]. Of course, it is also possible to extract the characteristics of theological interpretation by the quantitative analysis of distinctive and frequent vocabulary in the texts that describe the interpretation of the canon. It is also possible to extract the interpretation of the translator by comparing the correspondence between the original text and its translation into another language, because a translation is an interpretation of the original text [3].

In the second, direct approach, the extraction of characteristic words is fundamentally based on their frequency. Utilizing techniques such as TF-IDF, characteristic words can be extracted quantitatively. For characteristic vocabularies, it is possible to use co-occurrence [4] and dependency analyses to numerically clarify the usage tendencies of important words. Co-occurrence analysis is the study of word occurrences in common with the target word, and dependency analysis investigates the words in dependent relationships. The semantic analysis of words is fundamental to interpretation. These quantitative methods are equivalent to those used in the humanities, which are collectively termed concordance interpretation.

Another effective direct approach is to make comparisons between entire canonical texts or between small parts of canonical texts. This allows an analysis of the theological emphasis made when the canon was written [5] as well as a study of the process of creating the canon [6] on the basis of the quantitative comparison of several variant texts that are included in the canon. In addition, a comparison of parts of the text in the canon makes it possible to numerically verify the rhetorical structure that is constructed from the relationships among text parts [7].

In this paper, I would like to introduce four methods of quantitative analysis for the interpretation of religious thought in a scientific manner. The citation analysis of interpreters' texts and vocabulary analysis of translations can be considered as indirect approaches. For a direct approach, methods are introduced to extract canonical theology from differences in variant texts and to evaluate rhetorical structure. Moreover, I will discuss a software application for utilizing these algorithmic methods.

3 Co-citation Analysis of Religious Texts

3.1 Background

There are many theological differences between specific religious groups. As a result of these differences, interpretations of the canon can differ. If it is possible to scientifically extract these differences, the transition or mutual influence can be numerically analyzed.

Specifically, this method aims to automatically extract the main elements of a number of key conceptualizations from a religious text corpus and analyze their cluster construction using an objective and replicable methodology. This, in turn, will provide an objective basis for the examination of systematic thought [2].

3.2 Constructing Networks and Extracting Clusters

Here, we focus on the writings of St. Augustine and St. Thomas Aquinas, two influential Church Fathers, as well as those of Jean Calvin, Karl Barth, and Pope John Paul II. This enables us to extract essential teachings of Christian dogma through historical transitions and identify the individual characteristics of hermeneutics. Based on the patterns of Bible citations within their writings, we created networks for frequently cited sections of the Bible and extracted the main elements and clusters of these in order to compare a number of key conceptualizations. Clusters were extracted according to a threshold value of co-citation frequency. Table 1 gives the total number of citations and co-citations in each author's writings.

The resulting clustered network for Augustine is presented in Fig. 1. The nodes' alphabets and numbers are symbols that correspond to the Bible sections; dense parts are clusters. The differences in clusters extracted for each author are presented in Table 2.

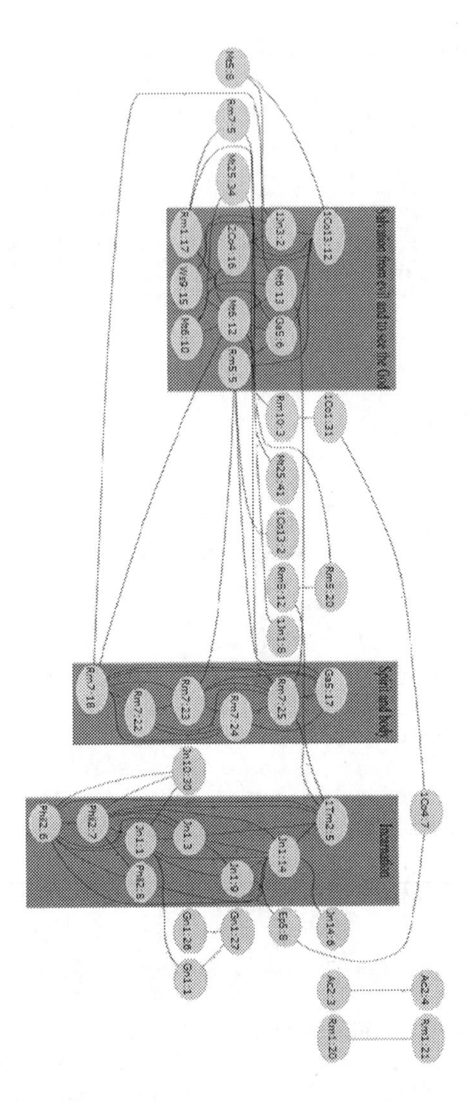

Fig. 1 Example of a Clustered Co-citation Network (Augustine)

Table 1 Citations and Co-citations

Author	Titles	Citations	Co-citations	Average citations per verse
Augustine	43	22674	215824	6.94
Thomas Aquinas	32	36015	800457	15.05
Jean Calvin	47	70324	2005864	13.51
Karl Barth	113	53288	2661090	23.67
John Paul II	1939	32166	643708	9.34

Table 2 Extracted Clusters

Cluster \ Author	Augustine	Thomas Aquinas	Jean Calvin	Karl Barth	John Paul II
Incarnation	○	○		○	○
Salvation from evil	○		○		
Spirit and body	○			○	
Predestination			○		
Commandments			○	○	
Evangelization				○	○
Sola Fidei				○	
Suffering servant				○	
Creation					○
Judgment					○

3.3 Discussions

This analysis identified the core element of Christian thought to be incarnation, because almost all the famous theologians shared the same cluster about incarnation (which includes Jn1:14, Phil2:6, Phil2:7, Phil 2:8, Gal4:4). In addition, distinctions between individual theologians in terms of their sect (Protestant theologians share a cluster about the Commandments) and era (modern-age theologians share a cluster about evangelization) were identifiable.

As Christianity literally believes that Jesus Christ is the Messiah, the result indicating the core element of Christianity seems to be valid. Moreover, as Protestants resist the rules of the Catholic Church, it is reasonable that they might emphasize the Commandments of the Bible instead of those of the Catholic Church. Likewise, in the modern age of science and globalization, modern churches need to strengthen the concept of evangelization.

The co-citation analysis results seem to match the circumstances of each theologian. This method could be applicable to other theological corpora.

4 Extracting the Interpretive Characteristics of Translations

4.1 Background

Although there have been some studies that focus on background interpretations by comparing and analyzing translations, these have utilized the methodologies of the humanities, which are unsuitable for maintaining objectivity and for large-scale analysis. Utilizing information technologies, this paper proposes some methods for numerical comparisons and the extraction of background interpretations in translations.

Specifically, the first step is to estimate the correspondence between the original vocabulary and the translation of that vocabulary on the basis of quantitative data. The next step is to objectively and quantitatively extract the differences in translators' interpretations from the differences in corresponding vocabulary in each translation [3].

4.2 Extracting Corresponding Vocabularies

Various high-performance algorithms are available for extracting corresponding word pairs from original and translated texts. These algorithms emphasize precision rather the recall ratio, because there are many large size corpora available for modern languages. However, in the case of classic texts such as the Bible (in ancient Greek and Hebrew), there are not enough original texts for large-scale quantitative analysis.

First, a new algorithm is designed to identify word pairs between the original text and the translated version. The algorithm incorporates three features: a word-for-word correspondence hypothesis, a recalculation of mutual information after the elimination of identified pairs, and an asymptotic threshold reduction. Through the combination of these features, recall rates improve by 20% compared to conventional methods and it is possible to extract multiple words corresponding to each of those in the word pairs.

Three Japanese translations of the Bible (*Colloquial Japanese*, *New Japanese*, and *The New Interconfessional Translation*) were analyzed using the proposed method, and vocabulary pairs of ancient Greek and Japanese were extracted.

4.3 Creating Networks on the Basis of Correspondences

In the next stage, two types of network are created on the basis of word correspondences, and the characteristics of translated words are extracted by calculating centrality values.

The network creation steps are depicted in Figs. 2–4. These identify the vocabulary correspondences (bipartite graph in Fig. 2), the relationship between words in the original languages (Fig. 3), and the relationships between words in the translated languages (Fig. 4).

Next, a centrality analysis (closeness, betweenness, and Bonacich) was applied to the extracted networks. The network centers reflect the conceptual center of the texts, because the central words signify that some concepts were more frequently used as an integrating concept or hypernym.

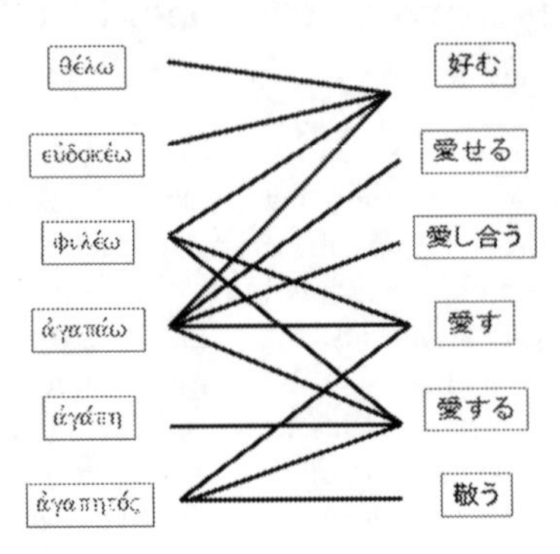

Fig. 2 Example of Corresponding Vocabularies

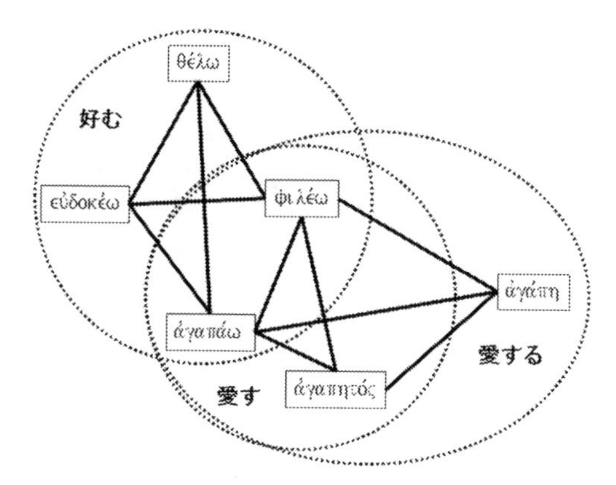

Fig. 3 Example of a Corresponding Network in the Original Language

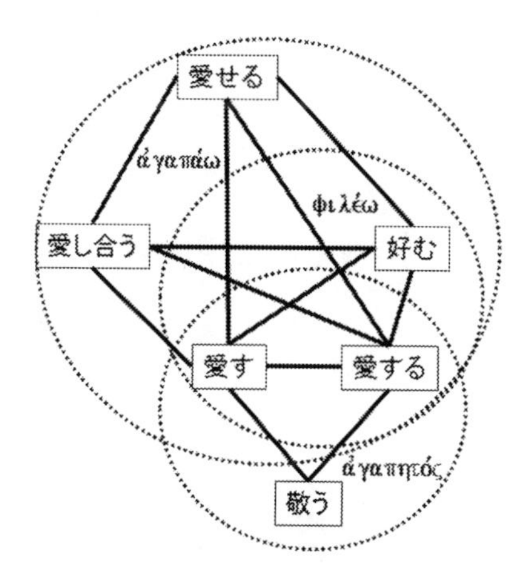

Fig. 4 Example of a Corresponding Network in the Translated Language

In addition to the relationships between original and translated languages, it is possible to make a network for the Bible that is composed of relationships between two original languages. The Old Testament was mainly written in Hebrew and the New Testament was mainly written in Greek. Therefore, the Bible describes the theology of the same God in two different languages. Because of this, modern Bible translations should interpret the conceptual theological relationships between Hebrew and Greek and translate them into one language. These relationships enable the correspondences between concepts in Hebrew and Greek to be analyzed using a modern translation as a medium.

As a case study, the relationship between the words *God* and *Lord* in the *New Revised Standard Version* (*NRSV*; Protestant translation) and *New American Bible* (*NAB*; Catholic translation) was analyzed. The results are depicted in Figs. 5 and 6, respectively.

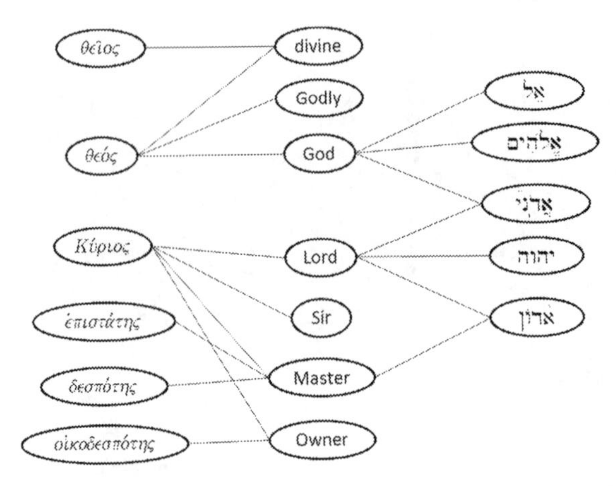

Fig. 5 Network of the conceptual relationship between *God* and *Lord* in *NRSV*

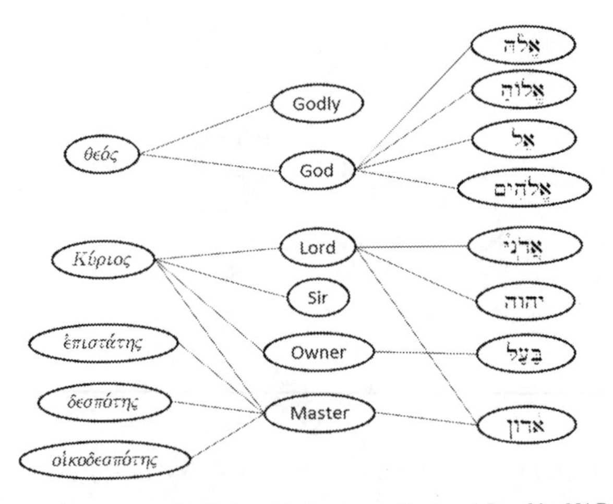

Fig. 6 Network of the conceptual relationship between *God* and *Lord* in *NAB*

4.4 Results and Discussion

The results from the vocabulary correspondence analysis of Japanese translations are summarized in Table 3. During translation, words that are important to a translator are translated carefully. Highly correspondent word pairs (in which concrete, mutual information is contained) between the original and translated text indicate characteristics of translation. Such highly correspondent word pairs were included in the results table.

The extracted results match the background of each translation. *Colloquial Japanese* is the oldest Japanese colloquial Bible. Fundamentalist Christians (equivalent to Evangelicals in the USA) were not satisfied with this version, and created the *New Japanese* version to emphasize the miracles and power of God. *The New Interconfessional Translation* was created to introduce a common Bible to the Catholic and Protestant churches, and therefore emphasizes peace among people of God.

Table 3 Characteristics of Translations

	Colloquial Japanese	**New Japanese**	**The New Interconfessional Translation**
Highly correspondent word	The Second Coming	Forgiveness and judgment on sins	Peace given by the Spirit, the Mission
Closeness centrality	Seeking for	*Fides quaerens intellectum*, the Mission	*Fides quaerens intellectum*
Betweenness centrality	Church as people who are called by God	Life is given by the Spirit	Coming of the Kingdom of God
Bonacich centrality	To know		

From the case study of conceptual relationships between two original languages, it can be inferred that *NRSV* interprets *God* and *Lord* as directly related concepts whereas *NAB* interprets them as fundamentally separated concepts. In other parts, word correspondences tend to be similar in the two translations. One Greek word was translated into several English words, and one English word was translated into several Hebrew words. Therefore, the level of detail in concepts about *God* and *Lord* is higher in Hebrew than in English or Greek.

From these results, it can be concluded that differences of interpretation between translations can be extracted by quantitative methods.

5 Synoptic Analysis of Religious Texts

5.1 Background

Undoubtedly, there are many cases where a group of people have sought to spread their message, and therefore developed a literature of *canonical documents*, but have encountered problems concerning the interpretation of the texts and the relationships between various individual documents. This kind of situation exists not only within Christianity, but also within other religions and schools of political thought. Such interpretative issues appear to have a direct influence on many matters in the modern world.

The central aim of this section is to develop a scientific information-technological method to analyze semantic differences that arise between multiple overlapping *canonical texts*. I believe that this method can be applied not just to the Bible, but also to the interpretation of systematic thinking embodied within collections of *canonical texts* in other spheres.

Specifically, this section introduces a method to analyze how central messages emerge from the existence of multiple overlapping *canonical texts*. This is applied to the four traditional Gospels in the Bible, allowing a comparison with the Catechism of the Catholic Church. This gives a numerical illustration of precisely which messages Christianity has sought to convey with the selection of the four traditional Gospels [5].

5.2 Creating Networks and Clusters

The internal structure of each Gospel is divided into segments called pericopes. *Pericope* is an ancient Greek word meaning *cut-out*. Each pericope corresponds to a small segment of a biblical story that was transmitted orally.
In the Gospels, pericope units are numbered, such as No. 235. However, a particular pericope in one Gospel may correspond to multiple pericopes in another Gospel. This one-to-many relationship is due to the editing process, as each Gospel writer combined pericopes that he believed were related. Thus, if one author saw a connection between one pericope and several others, that particular pericope unit would be repeated in a number of sections within the Gospel. Accordingly, there are many pericopes in the four Gospels that have the same verses, because they were taken and edited from the same source pericope.

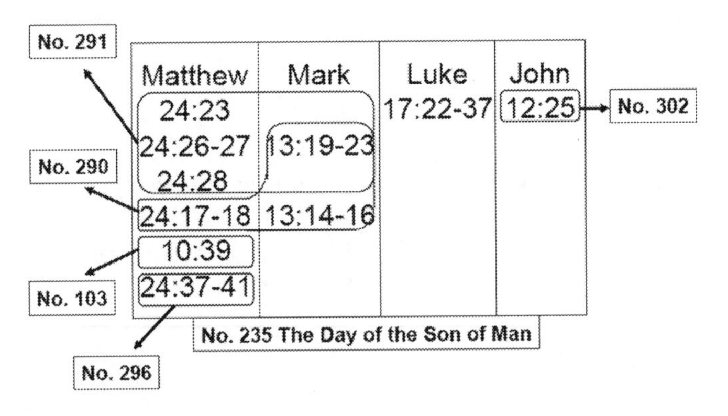

Fig. 7 Example of Pericope Relations

As Fig. 7 shows, pericopes containing verses in common with pericope No. 235 are Nos. 103, 290, 291, 296, and 302. This suggests that the writer of Matthew perceived some relationship among pericope Nos. 235, 103, 290, 291, and 296. Similarly, the writer of Mark imagined relationships between Nos. 290 and 291, whereas the writer of John made a link between pericope Nos. 235 and 302.

These pericope relationships can be converted into networks that regard pericopes as nodes and their relationships as edges. This study uses the *Synopsis of the Four Gospels* from the *Nestle-Aland Greek New Testament* (version 26) [8] as the data source of pericope relationships. This is believed to be the basis for various charts of pericope relations.

In order to identify the internal structure of the Gospels, the maximum connected subgraph was clustered and the core element was extracted. Four cores were extracted by combining node sharing cliques (Fig. 8 and Table 4). The following are the messages of the four cores:

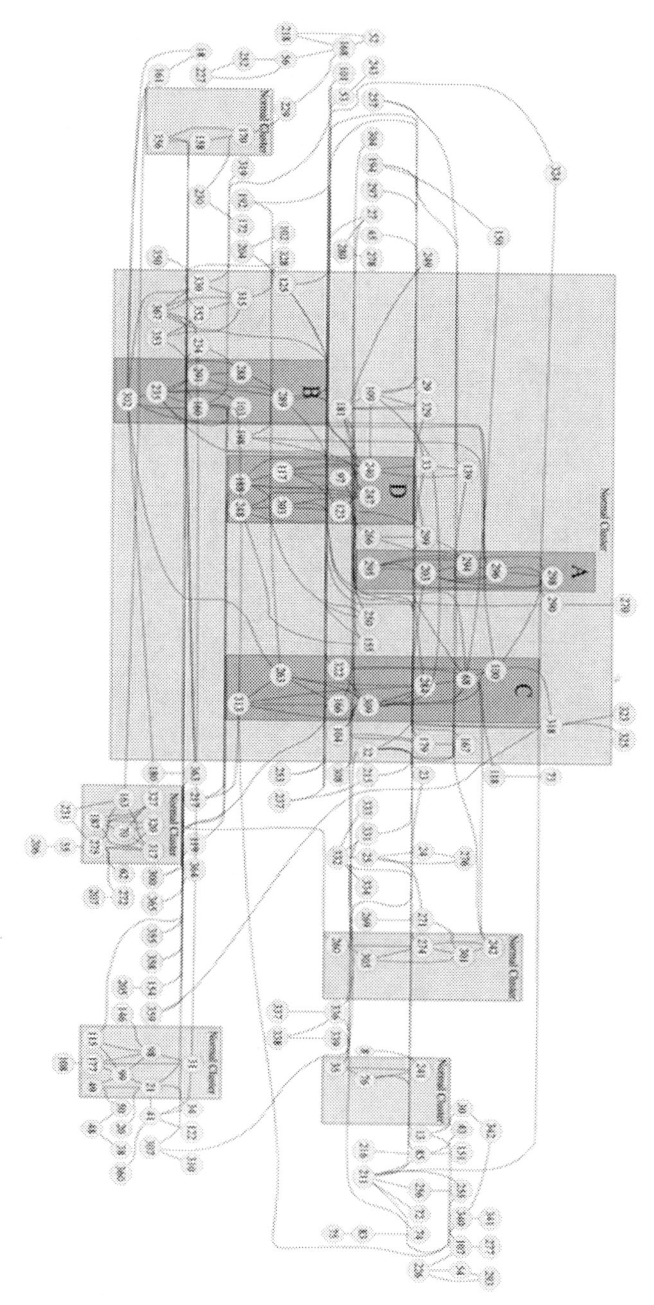

Fig. 8 Clustered Maximum Connected Partial Graph of the Four Gospels

Table 4 Contents of Clusters in the Network of the Four Gospels

Cluster	Pericope	Chapter and Verse of the Bible
A	203	Lk12:35–48
	294	Mk13:33–37
	295	Lk21:34–36
	296	Mt24:37–44
	298	Mt25:1–13
B	103	Mt10:37–39
	160	Mt16:24–28/Mk8:34–9:1/Lk9:23–27
	235	Lk17:22–37
	288	Mt24:3–8/Mk13:3–8/Lk21:7–11
	289	Mt24:9–14/Mk13:9–13/Lk21:12–19
	291	Mt24:23–28/Mk13:21–23
	302	Jn12:20–36
C	81	Lk6:37–42
	100	Mt10:17–25
	166	Mt18:1–5/Mk9:33–37/Lk9:46–48
	263	Mt20:20–28/Mk10:35–45
	284	Mt23:1–36/Mk12:37–40/Lk20:45–47
	309	Jn13:1–20
	313	Lk22:24–30
	322	Jn15:18–25
D	97	Mt9:32–34
	117	Mt12:22–30/Mk3:22–27
	188	Lk11:14–23
	240	Jn7:14–39
	247	Jn8:48-59

A) Preparation for the Day of Judgment because we do not know when it will come;
B) Foretelling persecution and recommending the path of discarding everything;
C) Teachings to the community of disciples;
D) Whether the miracles of Jesus were due to demons.

These teachings are believed to be the focus points of the old Church Fathers who canonized the New Testament.

In order to compare the results, the same analysis was applied to the *Catechism of the Catholic Church* [9]. Each item of the Catechism has a number, which is used as its ID. The ID numbers range from 1 to 2865. The relationships among the numbered items are complicated, and it is not unusual for one item to be related to several others. It is possible to construct a network by regarding items as nodes and relations as links, as for the pericopes in the Gospels.

As a result, ten clusters were extracted from the network of the Catechism's relationships (details of clusters are presented in Table 5). These clusters contained the following:

A) The Holy Spirit and the Sacraments;
B) The authorities of the Church;
C) The Virgin Mary;
D) The temptation of sin and malice; the miracles of the Christ;
E) Repentance, remittance, atonement;
F) Icons;
G) Human dignity in the figure of God;
H) Death;
I) Poverty;
J) Participation of laypeople in priesthood and prophecy.

Table 5 Contents of Clusters in Catechism

Cluster	Catechism ID Number
A	737, 788, 791, 798, 103, 1092, 1093, 1094, 1095, 1096, 1098, 1099, 1100, 1101, 1102, 1103, 1104, 1105, 1107, 1108, 1109, 1154
B	85, 86, 87, 88, 888, 889, 890, 891, 892, 2032, 2033, 2034, 2035, 2036, 2037, 2038, 2039, 2040
C	143, 148, 153, 485, 489, 494, 506, 722, 723, 726, 963, 1814, 2087, 2609, 2617
D	394, 518, 519, 538, 540, 542, 546, 550, 560, 1115, 2119, 2816, 2849
E	980, 1424, 1431, 1451, 1455, 1456, 1459, 1473
F	476, 1159, 1160, 1161, 1162, 2129
G	225, 356, 1700, 1703, 2258
H	958, 1032, 1371, 1689
I	544, 2443, 2544, 2546
J	784, 871, 901, 1268

5.3 *Discussion*

The themes of the three most concentrated clusters in the Catechism (corresponding to A, B, and C) are *the Holy Spirit and the Sacraments*, *the authorities of the Church*, and *the Virgin Mary*. These themes are typical of the Catholic Church, but are not approved by many Protestant churches. Considering that the emphasized messages are typically different in Protestant churches, it is possible that the differences between Catholics and Protestants were especially considered and enhanced when the Catechism was edited. Another typically Catholic characteristic is the cluster concerning icons (F).

The message that disciples should abandon everything to follow Jesus is included in cluster I, which has a size of 4; however, this is a small part of the entire network.

Other than that, the problems of liturgy (A) and authority in the Church (B) are closed up. In the four Gospels, *serve each other* is an important message; however, the Catechism insists upon the authority of priests. One main characteristic of the four Gospels is an eschatological warning, but the Catechism does not emphasize this. Overall, the messages to the disciples became more suitable for religious organization and the eschatological messages were weakened.

6 Validation Methodology for Classic Rhetorical Structure

6.1 Background

Literary criticism is a promising field for interpreting the Bible precisely. This methodology analyzes the Bible as literature and examines its use of literary techniques. A marked literary characteristic of the Bible is its sophisticated structures, which comprise classic rhetorical structures such as chiasmus (in Fig. 9), concentric structures (in Tables 6 and 7), and parallelisms.

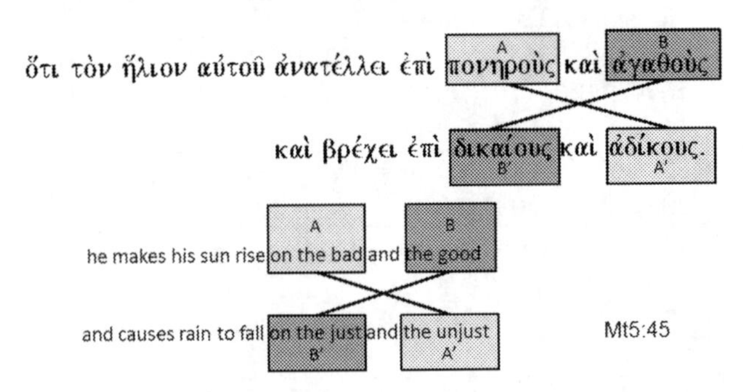

Fig. 9 Example of a Simple Chiasmus (Mt5:45)

Table 6 Example of Corresponding Pericopes in a Concentric Structure (Mk8:22–10:52)

	Part	**Name of Pericope**
A	8:22–26	Jesus cures a blind man at Bethsaida
B	8:27–30	Peter's declaration about Jesus
C	8:31–33	Jesus foretells his death and resurrection
D	8:34–38	Losing life for Jesus
E	9:1	A man who does not taste death
F	9:2–13	The transfiguration
G	9:14–29	The healing of a boy with a spirit
H	9:30–32	Jesus again foretells his death and resurrection
G'	9:33–50	Who is the greatest?
F'	10:1–12	Teaching about divorce
E'	10:13–16	Jesus blesses little children
D'	10:17–31	The rich man
C'	10:32–34	The third time Jesus foretells his death and resurrection
B'	10:35–45	The request of James and John
A'	10:46–52	The healing of blind Bartimaeus

Table 7 Example of a Corresponding Theme in a Concentric Structure (Mk8:22–10:52)

	Common Theme
A, A'	Healing the visually impaired
B, B'	Jesus is Messiah
C, C'	Foretelling death and resurrection
D, D'	Persecution and life
E, E'	Who enters the kingdom of the God
F, F'	Moses
G, G'	Evil spirit and child

There are several merits to identifying rhetorical structures in the Bible. It can clarify the divisions in a text; moreover, the correspondence of phrases signifies deeper interpretation. If the rhetorical structure is concentric, the main theme of that text is also clarified.

However, there are some problems regarding rhetorical structures. First, there is no clear definition of a valid correspondence. Some structures correspond by words or phrases, but more abstract themes may also be the element of correspondences. The length of the text unit is not uniform. Some structures are composed of phrases, whereas other structures are composed of pericopes. Therefore, a quantitative validation method for the rhetorical structure of the Bible is necessary [7].

6.2 *Evaluation Algorithm for Rhetorical Structures*

In this methodology, the relationships between each pericope in the rhetorical structure were first validated on the basis of the common occurrence of rare words and phrases. If corresponding pairs of pericopes more frequently include rare words and phrases, the probability of intentional arrangement is believed to be higher. Second, on the basis of the mean and standard deviation of a random combination of pericopes, the probability of accidental occurrences of common words and phrases in the test hypothesis is calculated. The common words and phrases are assumed to be normally distributed. The results are examined to determine whether the correspondences exceed the level of statistical significance.

Fig. 10 depicts an example calculation for the probability of random word pairs appearing when the rhetorical structure has nine pericopes and a particular word appears three times in that structure. The occurrence was counted in the form of a single word, a two-word phrase, and a five-word window; appearance thresholds of less than 10, 20, and 30% of pericopes were used for each form.

For the comparison, randomly divided pericopes were constructed from the same text. Two types of validation were also executed on the basis of the random division of pericopes. At first, the same combination of pericope patterns was applied to randomly divided pericopes. Next, a random combination of pericope patterns was applied to randomly divided pericopes. Thus, there are three estimates of validity:

A) Random combination and hypothetical pericope;
B) Hypothetical combination and random division;
C) Random combination and random division.

These three random situations were used to statistically validate the hypothesis of rhetorical structures. *Parallel Concentric Structures within the Bible* [10–12] was selected for the hypothesis. Tables 8–10 depict the results of the three types of validity estimation. The symbol ** signifies a 1% level of statistical significance, * signifies 5%, and + signifies 10%.

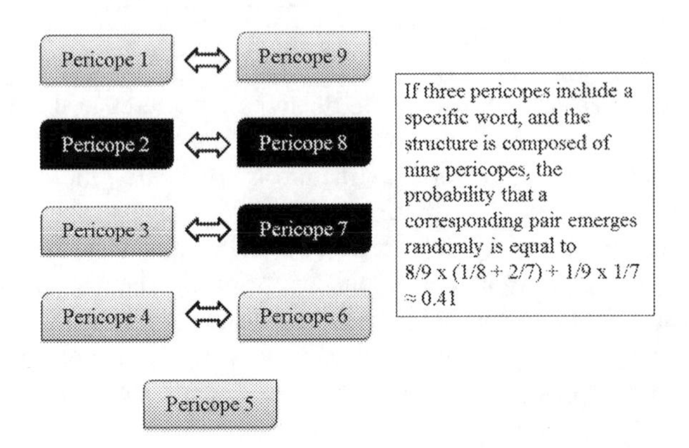

Fig. 10 Calculating the Probability of a Random Pair

Table 8 Statistical Evaluation of Rhetorical Structure 1 (Genesis-Deuteronomy)

		One word			**Two-word phrase**			**Five-word window**		
		10%	**20%**	**30%**	**10%**	**20%**	**30%**	**10%**	**20%**	**30%**
Genesis	**A**	*	**	**				**	**	**
	B	**		+						
	C	**		*						
Exodus	**A**			*	**	**	**			
	B		*	*	+	*	+	*	*	*
	C	+	**	*	*	*	*	**	*	*
Leviticus	**A**	+	*							
	B		*	+						
	C		**	*						
Numbers	**A**									
	B		**	+						
	C		**	*						
Deuteronomy	**A**	+	+							
	B					*	+		+	+
	C		+						+	*

Table 9 Statistical Evaluation of Rhetorical Structure 2 (Samuel-Ezekiel)

		One word			Two-word phrase			Five-word window		
		10%	20%	30%	10%	20%	30%	10%	20%	30%
Samuel 1, 2	A	*	*		**	*	**			
	B	*								
	C	*								
Kings 1, 2	A		**							
	B		*		*	*	**			+
	C		*		*	**	**			+
Isaiah	A	*			**	**	**			
	B						*			
	C				+		**			
Jeremiah	A	**	**	**	*					
	B		**							
	C		**			+			+	
Ezekiel	A								+	*
	B								*	+
	C							+	*	*

Table 10 Statistical Evaluation of Rhetorical Structure 2 (Matthew-Revelation)

		One word			Two-word phrase			Five-word window		
		10%	20%	30%	10%	20%	30%	10%	20%	30%
Matthew	A			*				*		
	B		+		+			*	**	**
	C							*	*	**
Mark	A	*	**	**						
	B	+	**	*						*
	C	+	**	*						
Luke	A	+	*							
	B									
	C	+								
John	A	*								
	B									
	C	+								
Acts	A									
	B	*								
	C	*								
Revelations	A	**	**	*	**	**	**	**	**	*
	B	*	*	**	**	**	**	**	**	**
	C	*	**	**	**	**	**	**	**	**

6.3 Results and Discussion

The results show that, in many texts of the Bible, the hypothesis is validated in terms of the relationships between words that occurred in less than 10% and 20% of the text, in the forms of either A or B or C.

A similar tendency in statistical significance among various books of the Bible seems to confirm that a unified rhetorical structure is included in these texts.

7 Software Application for Interpretation of the Bible

A project to develop a computer application to perform the above analyses is ongoing. The latest development version of a Java-based server–client model has been published [12].

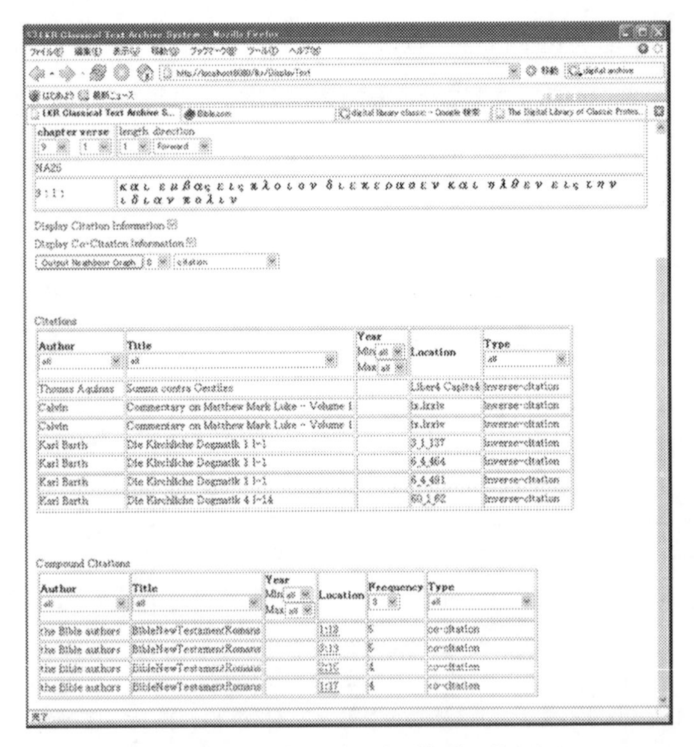

Fig. 11 Screenshot of the Interface for Browsing the Citation Database

The software contains an implementation of the algorithms described in this paper and data to support the interpretation. In addition, it includes general functions of Bible software, such as browsing and searching through several translations in parallel.

For the citation analysis function, a citation and reference database and browsing interface have been implemented (Fig. 11 shows a screenshot of the citations and references browsing interface), as has a network analysis function for the citation network. Furthermore, a browsing interface for cited text is included so that researchers can interpret the Bible on the basis of information about the relationships between citations and references (due to copyright issues, this function is not available in the test version).

For the translation analysis, an asymptotic correspondence vocabulary presumption method has been implemented as the estimation algorithm. This function outputs the results of vocabulary estimation in CSV format by dividing data on the original text and the translated text into morphemes. The next version of the software will enable users to browse the results of vocabulary estimation for each translated text alongside the original and translated texts.

An algorithm for verifying the validity of the rhetorical structure has also been implemented. In addition, the software includes a database of rhetorical structure hypotheses, and users can browse rhetorical structures for each text location (Fig. 12 shows a screenshot of the rhetorical structure browsing interface).

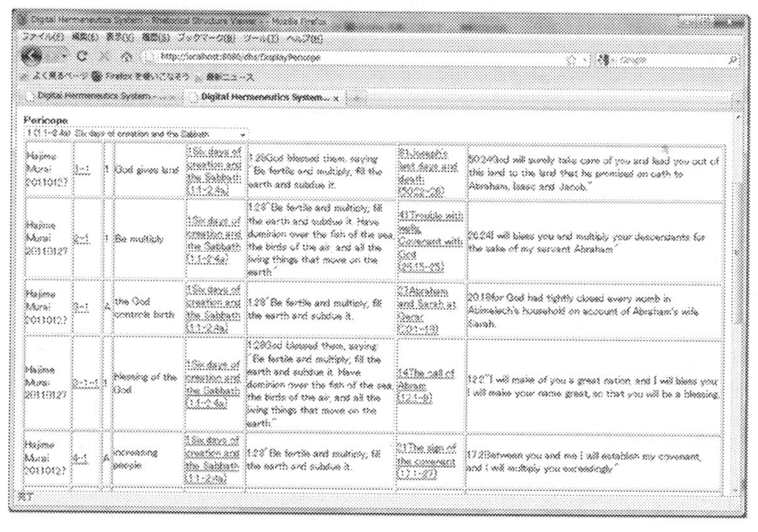

Fig. 12 Screenshot of the Interface for Browsing the Rhetorical Structure Database

8 Conclusions

This paper introduced four methods for the quantitative analysis of the interpretation of a canon of systematic thoughts. Although various quantitative methods can be used to analyze interpretations, it is difficult to analyze contextual information that is not described in texts (such as the circumstances of the author, historical facts, or cultural backgrounds). As in other fields where quantitative methods have not been applied, narratology and discourse analysis must also be used. To enable the scientific analysis of thought, it is necessary to resolve these problems and enrich the algorithms for the quantitative analysis of interpretation. These new algorithms should also be easily available to many researchers. Therefore, computer software supporting interpretation should implement an interface that allows an integrated analysis and a flexible combination of the results of several other algorithms.

Acknowledgments. This work was supported by KAKENHI 22700256 (Grant-in-Aid for Young Scientists [B]).

References

1. Murai, H., Tokosumi, A.: Extracting concepts from religious knowledge resources and constructing classic analysis systems. In: Tokunaga, T., Ortega, A. (eds.) LKR 2008. LNCS (LNAI), vol. 4938, pp. 51–58. Springer, Heidelberg (2008)
2. Murai, H., Tokosumi, A.: Co-citation network analysis of religious text. Transactions of the Japanese Society for Artificial Intelligence 21(6), 473–481 (2006)
3. Murai, H.: Extracting the interpretive characteristics of translations based on the asymptotic correspondence vocabulary presumption method: Quantitative comparisons of Japanese translations of the Bible. Journal of Japan Society of Information and Knowledge 20(3), 293–310 (2010)
4. Landauer, T.K., Foltz, P.W., Laham, D.: Introduction to latent semantic analysis. Discourse Processes 25, 259–284 (1998)
5. Murai, H., Tokosumi, A.: Network analysis of the four Gospels and the Catechism of the Catholic Church. Journal of Advanced Computational Intelligence and Intelligent Informatics 11(7), 772–779 (2007)
6. Miyake, M., Akama, H., Sato, M., Nakagawa, M., Makoshi, N.: Tele-synopsis for biblical research: Development of NLP based synoptic software for text analysis as a mediator of educational technology and knowledge discovery. In: Conference on Educational Technology in Cultural Context (ETCC) in conjunction with ICALT, pp. 931–935 (2004)
7. Murai, H.: A study about validation methodology for classic rhetorical structure by utilizing digital archive. In: IPSJ SIG Computers and the Humanities Symposium 2011, vol. 8, pp. 211–218 (2011)
8. Kurt, A.: Synopsis of the four Gospels revised standard version. The United Bible Societies (1982)
9. Paul II, J.: Catechism of the Catholic Church (1992), `http://www.vatican.va/archive/ccc/index.htm`
10. Murai, H.: Parallel concentric structure within the Gospel of Mark. Journal of Catholic Theological Society of Japan 20, 65–95 (2009)
11. Murai, H.: The parallel concentric structures within Exodus. Society of Biblical Literature Annual Meeting (2010).
12. Murai, H.: Rhetorical Structure of the Bible (2012), `http://www.valdes.titech.ac.jp/~h_murai/`

Semantic Annotation of Web Services: A Comparative Study

Djelloul Bouchiha, Mimoun Malki, Djihad Djaa,
Abdullah Alghamdi, and Khalid Alnafjan

Abstract. A Web service is software that provides its functionality through the Web using a common set of technologies, including SOAP, WSDL and UDDI. This allows access to software components residing on different platforms and written in different programming languages. However, several spots, including the service discovery and composition, remain difficult to be automated. Thus, a new technology has emerged to help automate these tasks ; it is the Semantic Web Services (SWS). One solution to the engineering of SWS is the annotation. In this paper, an approach for annotating Web services is presented. The approach consists of two processes, namely the categorization and matching. Both processes use ontology matching techniques. In particular, the two processes use similarity measures between entities, strategies for computing similarities between sets and a threshold corresponding to the accuracy. Thus, an internal comparative study has been done to answer the questions: which strategy is appropriate to this approach? Which measure gives best results? And which threshold is optimum for the selected measure and strategy? An external comparative study is also useful to prove the efficacy of this approach compared to existing annotation approaches.

Keywords: Annotation, Web Service, SAWSDL, Semantic Web Services, Ontology Matching.

Djelloul Bouchiha · Mimoun Malki
EEDIS Laboratory, Djillali Liabes University of Sidi Bel Abbes, Algeria
e-mail: `bouchiha.dj@gmail.com`, `malki@univ-sba.dz`

Djihad Djaa
University Dr. Taher Moulay, Computer Department, Saida, Algeria
e-mail: `djihad22@hotmail.fr`

Abdullah Alghamdi · Khalid Alnafjan
College of Computer and Information Sciences. KSU, Riyadh, Saudi Arabia
e-mail: `{Ghamdi,Alnafjan}@KSU.EDU.SA`

R. Lee (Ed.): *SNPD*, SCI 492, pp. 87–100.
DOI: 10.1007/978-3-319-00738-0_7

1 Introduction

The Web services technology allows accessing to heterogeneous software, in terms of languages and platforms, through the Web with common standards, including SOAP[1], WSDL[2] and UDDI[3]. However, the syntactic nature of these standards has hindered the discovery and composition of these services. To resolve this problem, semantic Web services have emerged. To add semantics to a service, it is possible to annotate the elements of this service with the concepts of an existing domain ontology. The annotation consists to associate the WSDL elements of a Web service with concepts of an existing semantic model. Often this model is a domain ontology of the Web service.

In preliminary work, an annotation approach has been proposed in [3]. It consists of two main processes: categorization which classifies the WSDL document in its corresponding domain, and matching which associates each entity of the WSDL document with the corresponding entity in the domain ontology. Both categorization and matching are based on ontology matching techniques [7] which in turn use similarity measures between entities. A similarity measure quantifies how much two entities are similar. In particular, WordNet based similarity measures are used [17].

To compare the results from the annotation approach using different similarity measures (internal comparison), and compare this approach with other existing approaches (external comparison), a tool called SAWSDL Generator has been implemented. The tool receives as input a WSDL file and a set of domain ontologies, and then generates a WSDL document annotated according to the SAWSDL standard [8].

This paper presents an internal comparative study to improve, optimize and determine under which conditions the annotation approach provides its best results. An external comparative study is also presented to show clearly the effectiveness of this approach over other annotation approaches.

The paper is organized as follows: the next section presents a summary of the approaches of Web services annotation which exist in the literature. Section 3 presents the annotation approach presented in [3]. Section 4 and Section 5 detail a comparative study. The final section concludes our work while presenting the main contributions that the comparative study enabled us to provide.

2 Literature Review

The annotation of a Web service consists in associating and tagging WDSL elements of this service with the concepts of an ontology [16].

---

[1] http://www.w3.org/TR/soap/
[2] http://www.w3.org/TR/wsdl
[3] http://www.uddi.org

Several approaches have been proposed for annotating Web services. Table 1 summarizes the characteristics of the Web service annotation approaches as follow: (1) The "Approach" column corresponds to the approach in question ; (2) The "Considered elements" column describes the considered elements in the annotation process ; (3) The "Annotation resource" column indicates the model from which semantic annotations are extracted ; (4) The "Techniques" column presents the used techniques for the annotation ; (5) The "Tool" column indicates the tool supporting the approach.

Table 1 Summary of Web service annotation approaches

Approach	Considered elements	Annotation resource	Techniques	Tool
[2]	Operation parameters	Workflow	Parameter compatibility rules	Annotation Editor
[3]	Complex types and operations names	Domain ontology	Ontology matching	SAWSDL Generator
[10]	Operations, message, parts and Data	Domain ontology	Text classification techniques	ASSAM
[16]	Data (Inputs and Outputs of services)	Domain ontology	Schema matching techniques	MWSAF tool
[9]	Natural-language query	Domain Ontology	Text mining techniques	Visual OntoBridge (VOB)
[13]	Data (Inputs and Outputs of services)	Meta-data (WSDL)	Machine learning techniques	Semantic labelling tool
[4]	Annotation & Query	Workflow	Propagation method	Prolog Implementation
[5]	Datalog definitions	Source definitions	Inductive logic search	EIDOS

There are also many other tools in semantic annotation like CharaParser which is a software application for semantic annotation of morphological descriptions [6]. Jonquet et al. [11] developed NCBO Annotator, an ontology-based web service for annotation of textual biomedical data with biomedical ontology concepts. SemAF (Semantic annotation framework) allows the specification of an annotation language [18]. Wyner and Peters [19] propose several instances of GATE's Teamware to support annotation tasks for legal rules, case factors and case decision elements. Liao et al. [14] identifie three main components of semantic annotation, propose for it a formal definition and presents a survey of current semantic annotation methods.

In a preliminary work, bouchiha et al. [3] propose to annotate Web services using ontology matching techniques. In the rest of this document, we usually use the name of the tool "SAWSDL Generator" to reference this approach.

3 SAWSDL Generator

As shown in Fig. 1, the annotation approach consists of two main processes: categorization and matching. Both categorization and matching are based on ontology matching techniques. The goal of ontology matching is to find relationships between entities [7]. Usually, these relations are equivalences discovered through similarity measures computed between these entities. To be accomplished, the ontology matching process uses similarity measures between entities. A similarity measure quantifies how much two entities are similar [17].

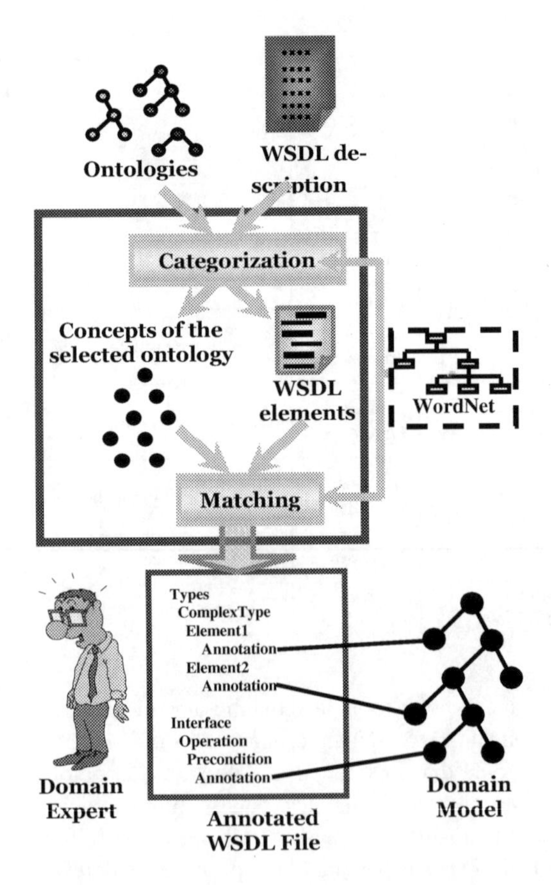

Fig. 1 SAWSDL Generator Architecture [3]

SAWSDL Generator uses similarity measures based on WordNet, such as Path, Resnik, Lin and Jiang [17]. WordNet is an online lexical database designed for use in a program [15]. Thus, these measures are calculated and then normalized. Normalization usually involves reversing the measured value to obtain a new value between 0 and 1. The value 1 indicates a total semantic equivalence between the two entities.

When multiple ontologies are available, the similarity between sets must be calculated by comparing the set of entities of the WSDL file with all entities of each ontology. Based on these measures, the system chooses one among ontologies to run the matching algorithm. The selected domain ontology determines the category of the WSDL document. This process is called "Categorization Process".

SAWSDL Generator treats an ontology as a set of entities (concepts), and a WSDL document also as a set of entities (XSD data types, interfaces, operations and messages). In data analysis, the linkage aggregation methods, including Single Linkage, Full Linkage and Average Linkage, allow the assessment of the distance between two sets of elements [7].

After the categorization, the elements of the WSDL document are associated to the ontology concepts using a similarity measure. This process is called "Matching Process".

The whole annotation process is a semi-automatic process where the user can intervene to validate, modify or reject the results. He can also set a threshold for the similarity measure. The threshold represents the required accuracy, it is a value between 0 and 1, the value 1 indicates that there must be a total semantic equivalence between two concepts. The use of a threshold reflects the allowable tolerance during calculations. More the threshold value is larger, more the results are accurate.

In the next section, a comparative study allows to evaluate and compare the results of the annotation process using different similarity measures.

4 Internal Comparison

The purpose of the internal comparison is to determine the measure, the strategy and the threshold that return the best results for the annotation process.

SAWSDL Generator offers the user the ability to choose the measure, the strategy and the threshold used in the categorization and matching. Thus, the comparative study is carried out for each one of two processes:

4.1 Categorization Evaluation

The categorization allows to associate a service to its corresponding domain. In this section, we use the term categorization or classification to say the same thing.

To check the categorization process with different similarity measures, we used a corpus[4] of 424 Web services organized manually into 26 areas (categories) [10]. Because of the lack of domain ontologies we limited our study to 40 services organized as follows: 14 *business* domain, 2 *travel*, 4 *money*, 8 *weather*, 6 *web*, 5 *matematics* and one service from the *music* domain. The domain ontologies used in the categorization process are *business*[5] and *travel*[6]. As a strategy for

---

[4] `http://www.andreas-hess.info/projects/annotator/ws2003.html`

[5] `http://www.getopt.org/ecimf/contrib/onto/REA/index.html`

[6] `http://protege.cim3.net/file/pub/ontologies/travel/travel.owl`

calculating similarity between sets we have chosen the *Average Linkage* where the similarity between two sets of entities is equal to the average similarity of all pairs of entities in both sets.

To evaluate and compare the results of the categorization process for different similarity measures, we use in the following the metrics precision, recall and F-measure [1].

- Recall (R): proportion of the correctly classified (categorized) services of all the services of the two considered domains, namely business and travel: R = CEN/CN.
- Precision (R): proportion of the correctly classified (categorized) services of the automatically classified services: P = CEN/EN.

The recall and the precision can be combined into a single measure, called F-measure [12], defined as follows: F-measure = (2 * recall * precision) / (recall + precision).

The following figure shows the curves of the F-measure of different similarity measures according to the threshold values:

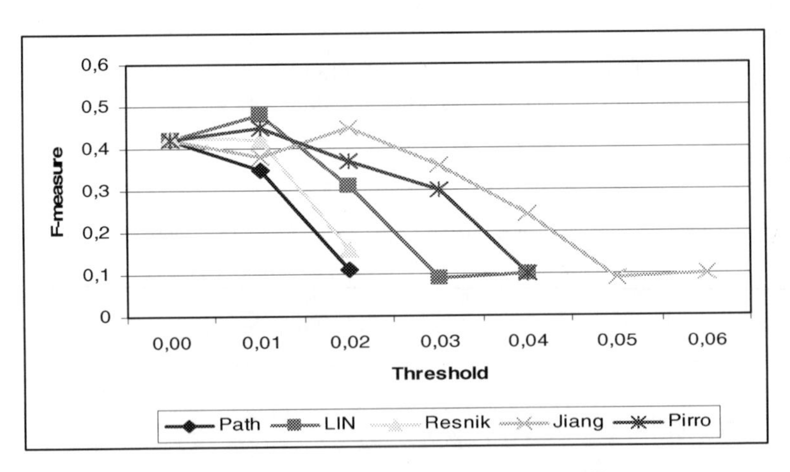

Fig. 2 The F-measure curves for the categorization process with different measures

The graph above shows that for a zero threshold, the different measures give the same results of categorization. However, the difference is between the scope of the different curves. This can be justified by the fact that certain measures return high average compared to others. For example, the *Jiang* measure gives higher averages compared to the *Path* and *Resnik* measures.

With more accuracy (threshold> 0) the results of categorization differ between these measures. The measure *Path* gives the lowest results of categorization compared to the other measures. We can see that with a threshold between 0.00 and

0.015, the *Lin* measure is the most effective for the categorization. For this interval, the measures *Resnik*, and *Pirro* provide best categorizations compared with the *Jiang* measure although it gives very high semantic averages. This may be justified by the fact that with the *Jiang* method several services out of domain are incorrectly classified; which is not the case with the other measures. However, for a threshold in the interval 0.015-0.02, the curves level changes so that the *Jiang* measure becomes the most powerful. When the threshold increases the *Jiang* measure classifies correctly most of the services; which is not the case for the measures *Resnik*, *Lin* and *Pirro*. For the same reasons, with a threshold between 0.00 and 0.015, the *Pirro* measure gives poor results compared to the *Lin* measure. With a threshold>0.015 the *Pirro* measure becomes better than the *Lin* measure but remains low compared to the *Jiang* measure.

We can deduce that if we use the measures *Lin*, *Resnik* and *Pirro* (usually measures that return low semantic averages), then the threshold must be between 0.00 and 0.015, to have best results. However, if we use the *Jiang* method, we must increase the threshold.

There is a dependency between the choice of the threshold and the choice of the measure. Note that for a threshold<0.015, it is recommended to use the *Lin* measure. However, with a threshold>0.015 it is recommended to use the *Jiang* measure. From the graph, we have also seen that the best classification has been obtained with the *Lin* measure, and with threshold=0.01.

The following table shows a ranking of measures based on the threshold depending on the quality of the results obtained by each measure:

Table 2 Ranking of the similarity measures with different thresholds

	0.00 – 0.01	**0.01 - 0.015**	**0.015 – 0.02**	**0.02 – 0.06**
1	Lin	Lin	Jiang	Jiang
2	Pirro	Pirro	Pirro	Pirro
3	Resnik	Jiang	Lin	Lin
4	Jiang	Resnik	Resnik	Resnik
5	Path	Path	Path	Path

A measure of lower rank is better compared to a measure of higher rank for the considered interval of threshold.

4.2 Matching Evaluation

We define the function *Match* which associates a WSDL element with the closest concept in the domain ontology. The function *Match* is defined as follows:

$$Match : E_{wsdl} \rightarrow E_{ontology}$$

$$e \rightarrow c, \textbf{ such as } sim(e,c) = MAX_{i=1}^{n} sim(e,c_i) \textbf{ and }$$

$$sim(e,c_i) \geq threshold$$

With E_{wsdl} the set of elements of the WSDL document, $E_{ontology}$ the set of concepts of the domain ontology, e is an element of E_{wsdl}, c is a concept of $E_{ontology}$, n the number of concepts of the domain ontology and *sim* a similarity measure.

To evaluate the matching process, we tested the performance of the association function *Match* with different similarity measures. For this, we have chosen a reference WSDL document, annotated manually, and consider it as valid. The selected WSDL document was *"TrackingAll"*. As a similarity strategy we have chosen *Average Linkage*, and as similarity measures we compared the measures *Path*, *Lin*, *Resnik*, *Jiang* and *Pirro*. To evaluate the results of the matching process, we used the metrics precision, recall and F-measure.

The following figure shows the curves of the F-measure of different similarity measures according to the threshold values:

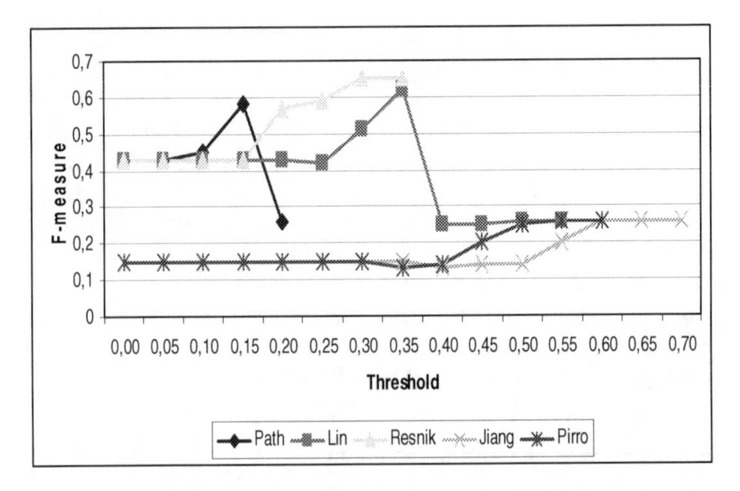

Fig. 3 The F-measure curves for the matching process with different measures

From this graph, we note that with a threshold belonging to 0.05-0.20, the measure *Path* performs best compared to the other measures.

The results of the measures *Jiang* and *Pirro* are low compared to the other measures, despite they return the highest similarities. This can be justified by the fact that with the measures *Jiang* and *Pirro*, all the elements containing the term *"Tracking"* are not correctly associated to their corresponding concepts, unlike *Path*, *Resnik* and *Lin*.

We note also that the curves of *Lin* and *Resnik* are similar in that they return the same results for a threshold belonging to 0.00-0.15. For a threshold belonging to 0.15-0.35, *Resnik* is more efficient compared to *Lin*. For a threshold> 0.35, *Resnik* returns no association.

It is therefore clear that the measures *Jiang* and *Pirro* return the lowest results for the matching process. With the measures *Lin*, *Path* and *Resnik*, the matching performance depends on the chosen threshold. It should be a compromise between the chosen threshold and the used similarity measure. With a threshold of 0.05 - 0.15, it is recommended to use the measure *Path*; the measures *Resnik* and *Lin* are ranked second.

4.3 Discussion

By analyzing the results of the categorization and matching in the same time, we found that for a given threshold, the quality of results is different for both processes. For example, the best results of matching are obtained with a threshold equal to 0.35. However, with this threshold, services are not classified in any category, i.e. the results of the categorization process are low for a threshold equal to 0.35. If using an appropriate threshold for the categorization process, then the quality of matching drops. This is justified by the fact that the similarity values used in the matching process are high compared to the semantic averages returned when using the strategy *Average Linkage* to accomplish the categorization process.

This problem is therefore due to the strategy *Average Linkage* used to calculate the semantic average between a WSDL document and an ontology. For example, the element *"company"* has a strong relationship with *"association"*, but has no similarity with the other concepts of the *business* ontology: with the measure *Path*, *sim(company, association)* = 0.25. With the other concepts, *sim(company, concept x)* = 0. The semantic average between the WSDL document and the ontology = 0.25/33 = 0.0075. It is therefore clear that the semantic average (0.0075) used for the categorization is too low compared to the similarity (0.25) used for the matching.

To resolve this problem, we have three solutions:

- Choose two different thresholds for the categorization and the matching.
- Use two different strategies for both processes.
- Define a new strategy that allows, with a single threshold, to find some compatibility between the categorization process and the matching process.

We chose the third option, and we proposed a hybrid strategy between ***Single Linkage*** and *Average Linkage*.

The new strategy is defined as follows:

Definition: Given a similarity function σ: O × O → R. The measure between two sets is a similarity function $\Delta : 2^O \times 2^O \rightarrow R$, such as:

$$\forall x, y \subseteq O, \Delta(x, y) = \frac{\sum_{i=1}^{|x|} MAX_{j=1}^{|y|} \sigma(e_i, e_j)}{|x|}, with(e_i, e_j) \in x * y$$

The following graph illustrates the results of the F-measure of the categorization and the matching process with the *Lin* measure and the new strategy:

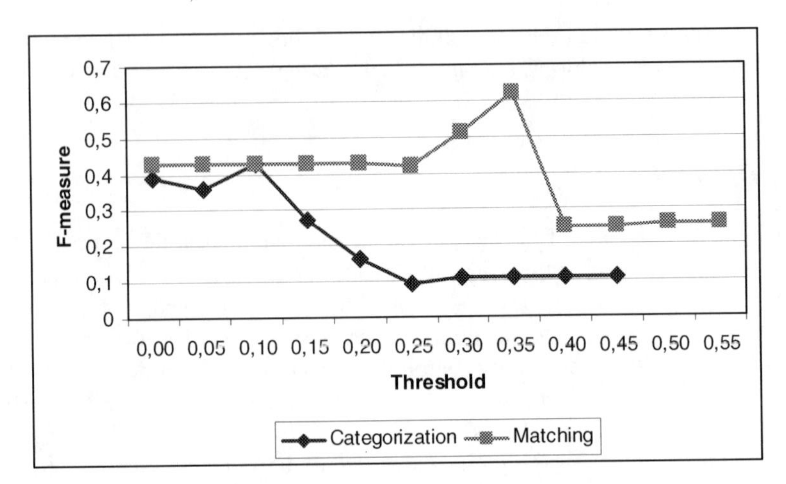

Fig. 4 The F-measure curves for the categorization and matching with the *Lin* measure

Finally, we can say that with the *Lin* measure, the new strategy and a threshold = 0.10, the annotation approach works best for the categorization and matching at the same time.

5 External Comparison

This section presents a comparative study between the approach presented in [3] and the other Web service annotation approaches which exist in the literature. Initially, our intention was to compare the approach presented in [3] with all other approaches. However, several problems have prevented us from making a complete study:

- Unavailability of technical details to implement these approaches.
- It is difficult to exploit the tools associated with these approaches.

Thus, we limited our study to the MWSAF approach [16]. MWSAF (METEOR-S Web Service Annotation Framework) is a Framework for semi-automatically annotate WSDL descriptions of Web services with relevant ontologies.

5.1 SAWSDL Generator vs. MWSAF

Purpose: The overall purpose of the two approaches is to semi-automatically annotate Web services with relevant ontologies. They consist of two phases: the categorization and matching. The categorization allows to classify a WSDL document in its corresponding domain. The matching allows to associate the elements of the WSDL document with their corresponding concepts in the ontology.

Used techniques: The approach MWSAF uses schema matching techniques. It uses the XML schema of the WSDL document and the schema of the ontology to compare them. SAWSDL Generator [3] uses ontology matching techniques by decomposing the WSDL document into its basic elements (XSD data types, interfaces, operations and messages) and the ontology into its concepts, then compares the two sets.

In both approaches, a similarity (or a semantic average) between the WSDL description and the ontology based on a similarity measure between two concepts is calculated to identify the domain ontology which will be retained for the annotation.

Two main differences can be distinguished:

- *The considered WSDL elements and ontology concepts:* To accomplish the annotation process, the two approaches compare the WSDL elements with the ontology concepts. MWSAF consider on the one hand the WSDL elements complexType, elementType, Enumeration and element, and on the other hand all the ontology concepts, namely class, sub-class, instance and attribute. SAWSDL Generator uses the WSDL elements complexType, elementType, element, operation, message and part, and it considers the class and sub-class as ontological concepts.
- *The similarity between two concepts:* While SAWSDL Generator uses linguistic similarity measures to compare two entities, MWSAF uses MS measure (match score) that is a combination between linguistic similarity and structural similarity where the similarity between two entities may depend on other concepts.

Since both approaches involve two processes, namely the categorization and matching, the comparison will focus on these two processes.

5.2 Comparison at the Categorization Level

To test the categorization process of the two approaches we have taken the same test base used for the internal comparison: 40 services including 16 of *business* domain, two of *travel*, 4 *money*, 8 *weather*, 6 *web*, 5 *matematics* and 1 service of *music*. The domain ontologies selected to apply the categorization process are *business* and *travel*. To evaluate the results of the categorization process we used the metrics precision, recall and F-measure.

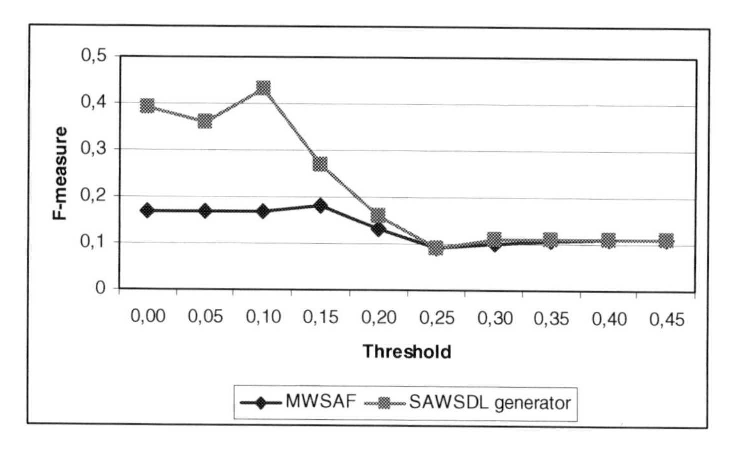

Fig. 5 The F-measure curves for the categorization process of the two approaches

From Fig. 5, we note that for a threshold>=0.25, both approaches give the same results of categorization. But with a threshold<0.25 the MWSAF approach gives results less efficient compared to SAWSDL Generator. The best results of SAWSDL Generator are obtained with a threshold=0.1.

5.3 *Comparison at the Matching Level*

To compare the matching results of SAWSDL Generator and MWSAF approach, we tested the performance of the association function according to the used similarity measure. Recall that the main difference between the two approaches is that SAWSDL Generator uses a linguistic similarity measure based on WordNet, and the MWSAF approach uses the measure MS (match score) which is composed of a linguistic similarity measure and structural similarity between two concepts, which relies on the similarity of sub-concepts.

For testing, we chose a reference annotated WSDL document and considered it as valid. The chosen WSDL document was *"TrackingAll"*. As similarity measure, we chose the *Lin* measure for SAWSDL Generator, and for the MWSAF approach we used the linguistic similarity measure *Ngram*. To evaluate the results of the matching process, we used the metrics recall, precision and F-measure.

The following figure shows the curves of the F-measure of the two approaches according to the threshold values:

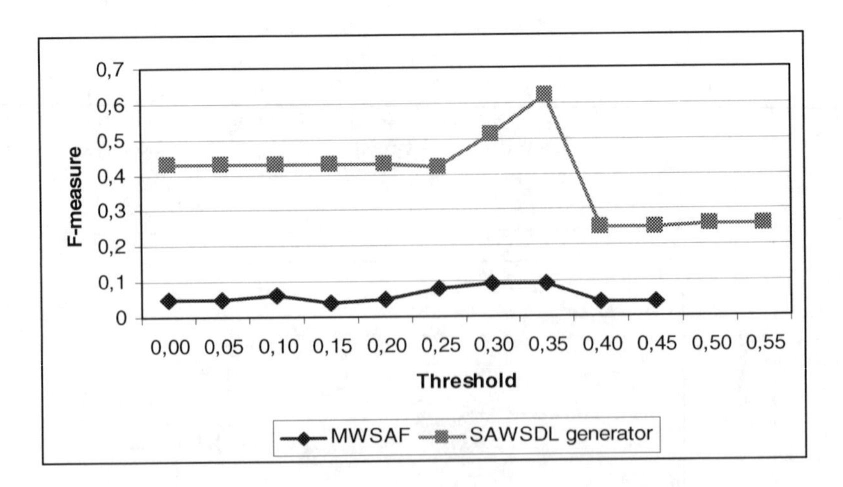

Fig. 6 The F-measure curves for the matching process of the two approaches

From this graph we see that the results obtained by MWSAF are underperforming compared to SAWSDL Generator whatever the chosen threshold.

6 Conclusion and Perspectives

To take advantage of Web service technology, an approach has been proposed to annotate syntactic WSDL descriptions of Web services by ontological models [3]. The annotation approach has two main processes: categorization and matching. In the first process, the WSDL description of a service is assigned to its corresponding domain. In the second process, the WSDL entities are associated with pre-existing domain ontology concepts. Both processes categorization and matching use similarity measures based on WordNet. A tool, called SAWSDL Generator has been developed to implement the proposed approach.

A comparative study has been conducted to improve the annotation approach and show its effectiveness compared to other works. The evaluation experiments showed that this approach works best with the *Lin* measure and a threshold equal to 0.1. To create a compromise between the categorization and matching, a new strategy for calculating similarity between sets has been proposed.

SAWSDL Generator, as it stands, provides very satisfactory and encouraging results. However, it still suffers from several problems including the problem of compound words and abbreviated words used as identifiers in the descriptions of services. While these are significant identifiers for service developers, they can not be handled by the Java WordNet API[7]. Hence the need for linguistic analysis for the separation between the composed parts in a word and identify the important parts (the parts that are meaningful to the domain), and the need to analyze the abbreviations. Until these problems are solved, human intervention remains necessary to correct the incorrect matches.

References

1. Baeze-Yates, R., Ribeiro-Neto, B.: Modern information retrieval. Addison-Wesley, ACM Press, Reading, MA (1999)
2. Belhajjame, K., Embury, S.M., Paton, N.W., Stevens, R., Goble, C.A.: Automatic annotation of web services based on workflow definitions. ACM Transactions on the Web (TWEB Journal) 2(2) (2008)
3. Bouchiha, D., Malki, M.: Semantic Annotation of Web Services. In: 4th International conference on Web and Information Technologies (ICWIT 2012), SBA Algeria, April 29-30 (2012)
4. Bowers, S., Ludäscher, B.: A calculus for propagating semantic annotations through scientific workflow queries. In: Query Languages and Query Processing workshop (QLQP-2006) anised in conjunction with the 10th International Conference on Extending abase Technology, pp. 712–723 (2006)
5. Carman, M.J., Knoblock, C.A.: Learning Semantic Definitions of Online Information Sources. Journal of Artificial Intelligence Research 30, 1–50 (2007)

[7] http://jwordnet.sourceforge.net/handbook.html

6. Cui, H.: CharaParser for fine-grained semantic annotation of organism morphological descriptions. Journal of the American Society for Information Science and Technology 63(4), 738–754 (2012)
7. Euzenat, J., Shvaiko, P.: Ontology Matching. English book. Springer, Heidelberg (2007)
8. Farrell, J., Lausen, H.: Semantic Annotations for WSDL and XML Schema. W3C Recommendation (2007), http://www.w3.org/TR/sawsdl/ (accessed August 28, 2007)
9. Grcar, M., Mladenic, D.: Visual OntoBridge: Semi-automatic Semantic Annotation Software. In: Buntine, W., Grobelnik, M., Mladenić, D., Shawe-Taylor, J. (eds.) ECML PKDD 2009, Part II. LNCS, vol. 5782, pp. 726–729. Springer, Heidelberg (2009)
10. Hess, A., Johnston, E., Kushmerick, N.: ASSAM: A Tool for Semi-Automatically Annotating Semantic Web Services. In: International Semantic Web Conference, Hiroshima Japan, pp. 320–335 (2004)
11. Jonquet, C., Shah, N., Youn, C., Musen, M., Callendar, C., Storey, M.: NCBO Annotator: Semantic Annotation of Biomedical Data. In: 8th International Semantic Web Conference (ISWC 2009) Posters and Demonstrations, Washington DC, USA (2009)
12. Larsen, B., Aone, C.: Fast and effective text mining using lineartime document clustering. In: Proceedings of the 5th ACM SIGKDD International Conference on Knowledge Discovery and Data Mining, pp. 16–22 (1999)
13. Lerman, K., Plangprasopchok, A., Knoblock, C.A.: Automatically labeling the inputs and outputs of web services. In: Proceedings of the National Conference on Artificial Intelligence (AAAI 2006), Boston, Massachusetts, USA (2006)
14. Liao, Y., Lezoche, M., Panetto, H., Boudjlida, N.: Semantic Annotation Model Definition for Systems Interoperability. In: The 6th International Workshop on Enterprise Integration, Interoperability and Networking (EI2N), Hersonissos Crete, Greece (2011)
15. Miller, G.A.: WordNet: An on-line lexical database. International Journal of Lexicography, 235–312 (1990)
16. Patil, A., Oundhakar, S., Sheth, A., Verma, K.: METEOR-S Web Service Annotation Framework. In: WWW 2004, pp. 553–562. ACM Press (2004)
17. Pedersen, T., Patwardhan, S., Michelizzi, J.: WordNet:Similarity - Measuring the Relatedness of Concepts. In: Proceedings of the Nineteenth National Conference on Artificial Intelligence (AAAI 2004), pp. 1024–1025 (2004)
18. Pustejovsky, J., Lee, K., Bunt, H., Romary, L.: ISO-TimeML: An International Standard for Semantic Annotation. In: Proceedings LREC 2010, La Valette Malte (2010)
19. Wyner, A., Peters, W.: Semantic Annotations for Legal Text Processing using GATE Teamware. In: The 4th Workshop on Semantic Processing of Legal Texts (SPLeT 2012) in Istanbul Turkey (2012)

VC-Bench: A Video Coding Benchmark Suite for Evaluation of Processor Capability

Xulong Tang, Hong An, Gongjin Sun, and Dongrui Fan

Abstract. Video coding standards such as H.264 are widely used since the flourish of digital compressed video ranging from low bit-rate Internet streaming applications to HDTV broadcast. As a result, more powerful processors are required to meet the growth in coding quality and velocity. Though many efforts have been made on associated items, it lacks one up-to-date benchmark suite for studying architectural properties of video coding applications. This paper introduces VC-Bench, a Video Coding Benchmark suite which is built up from a wide range of video coding applications. Firstly, typical codecs are selected (X264, XVID, VP8) according to the popularity, coding efficiency, compression quality, and source accessibility. Secondly, hotspots are extracted from coding process as kernels. VC-Bench mainly focuses on Transformation, Quantization and Loop filter. All of the extracted kernels are single-threaded version without any architecture-specific optimizations, such as SIMD. Besides, inputs in three different sizes are provided to control running time. Finally, to better understand intrinsic characteristics of video coding application, we analyze both computational and memory characteristics, and further provide insights into architectural design which can improve the performance of this kind of applications.

Keywords: VC-bench, video codec, hotspots, performance, architecture.

Xulong Tang · Hong An · Gongjin Sun
Department of Computer Science and Technology, USTC, Hefei,
Anhui, 230027, P.R. China
e-mail: `{tangxl,hilcutz}@mail.ustc.edu.cn, han@ustc.edu.cn`

Dongrui Fan
Institute of Computing Technology, Chinese Academy of Sciences, Beijing,100190, China
e-mail: `fandr@ict.ac.cn`

R. Lee (Ed.): *SNPD*, SCI 492, pp. 101–116.
DOI: 10.1007/978-3-319-00738-0_8

1 Introduction

In the past few decades, video technology was prevalent in many domains like communication, computer, radio and TV. Two significant trends are becoming clear. First, Applications like video conference, videophone and digital TV increase rapidly. This trend leads to the development of video coding standards, such as H264 and webM [1, 2]. A variety of codecs have been published according to these standards. Thus, it is paramount to ensure both coding quality and velocity to meet the demand of users. Second, though many powerful general-purpose processors have already been designed, simply increasing the number of cores does not automatically yield performance due to the complexity of video coding applications. Architectural parameters such as cache hierarchy and branch prediction mechanism also affect video coding performance. How to guarantee coding quality and velocity on devices such as multi-core processors, embedded platforms and mobiles become a cutting edge research topic.

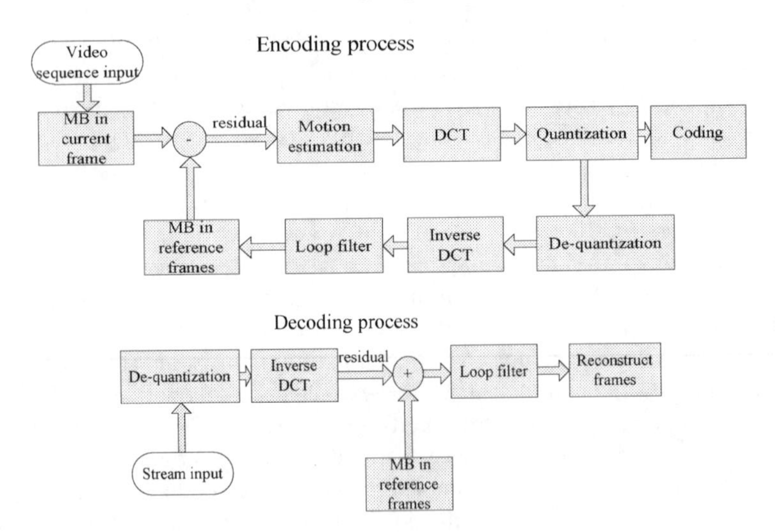

Fig. 1 Overview of video coding process

Fig. 1 shows an overview of video coding process. Encoding begins with parsing the input commands and dividing video sequence into consecutive frames. Each frame further splits into blocks called macroblock (MB) which contains 16X16 pixels. MB then goes through five basic phases called motion estimation (ME), discrete cosine transform (DCT), quantization, loop filter and entropy coding. Both encoding and decoding apply inverse transformation and de-quantization processes to construct reference frames. Each phase has its specific characteristics. For example, ME uses search algorithms to do block matching. This is an iterative process to find the most similar MB compared to current MB. It uses several frames as input and the majority of operations are computational instructions. DCT does transformation on MB while quantization seeks for high compression rate by reducing the precision of transformed data. Operations of these two phases are relatively simple with many regular memory accesses. Different program behaviors need different architecture support. Besides, today's prevalent performance analyzing tools cannot gather properties of each phase respectively. They provide overall results so that it is hard to analyze bottlenecks precisely. In order to evaluate processor capability and guide microarchitecture design, a benchmark suite concentrating on each phase is urgently needed.

In this paper, we present Video Coding Benchmark suite (VC-Bench) which is built up from widespread video codecs. VC-Bench includes codecs' hotspots which are the most time consuming parts. This benchmark suite is written in 'clean' C in order to provide a fair evaluation result on various architectures. Inputs from three resolutions are provided so that researchers can control simulation time and understand the influence of input size on performance. The specific contributions of this work are as follows:

- Introduce a new video coding benchmark suite, including kernels from video coding applications. These applications cover the widely used codecs in multimedia domain.
- Hotspots of selected codecs are analyzed. Further extraction of hotspots facilitates analysis on architecture performance.
- Three different sizes of input are offered to control running time. These inputs cover typical characteristics of real-world video sequences.
- Architectural properties are analyzed. Insights into architectural design are derived to improve the performance of video coding applications.

The rest of this paper is organized as follows. Section 2 introduces related work and section 3 presents the approach we use to build up VC-Bench. We provide the performance comparison and execution characteristics in section 4. In section 5, we draw a conclusion and discuss future work.

2 Related Work

There are many notable benchmark suites currently. MEVBench, building up from mobile embedded vision applications, was presented to guide embedded architecture design [3]. The author concentrated on key algorithms within mobile vision applications and analyzed dynamic code hotspots, computing properties, and memory system performance. San Diego Vision Benchmark Suite (SD-VBS) was provided by a team from University of California, San Diego [4]. It included diverse computer vision applications drawn from vision domain. Nine applications were listed and each contained several computational kernels. Same as MEVBench, they analyzed the hotspots and execution properties of selected applications.

Mediabench [5] was an advanced benchmark suite targeting at video coding. The author compared video coding applications with SPECint benchmarks and pointed out their differences on IPC and bandwidth. There were three generations of Mediabench. The latest one was published in 2006 [6]. Though it implemented H.264 and MPEG-4, we find Mediabench does not match the developing speed of video coding application. Firstly, limited video coding standards were included. Recently burgeoning standards such as webM [2] presented by Google were not included. Secondly, the codecs were out of style. They chose codec JM 10.2 [7] and FFMPEG [8] to represent H.264 and MPEG-4 codecs. But the most prevalent codecs of H.264 and MPEG-4 are X264 [9] and XVID [10] respectively. Thirdly, the memory behavior of video coding changed. In Fig. 2 we compare modern codecs with Mediabench and it is clear that they are different. It is important to use up-to-date applications to evaluate modern computer architecture.

A team from Barcelona Supercomputing Center-CNS, Spain published a benchmark suite called HD-VideoBench, which overcame shortcomings of Mediabench by including X264 and XVID and filled the blank of missing HD digital video [11]. However, they did not address on architectural performance which is usually the purpose to build a benchmark suite. Besides, HD-VideoBench did not analyze hotspots but chose full codecs instead. This approach results in long running time and various redundant codes. It is difficult to analyze characteristics of program behavior and to locate bottlenecks precisely.

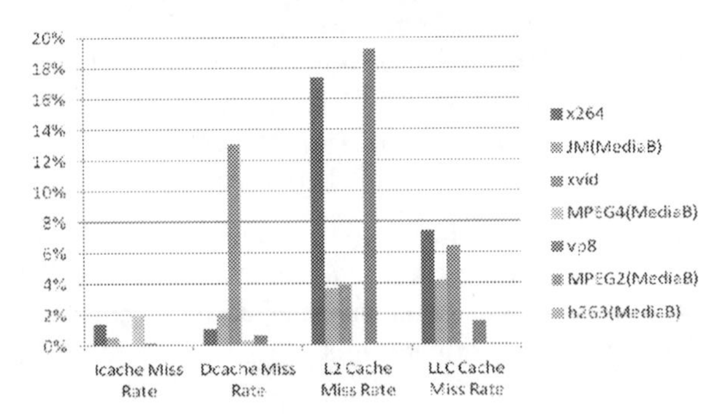

Fig. 2 Comparison of Mediabench and modern codecs

3 Details of VC-Bench

In order to build a comprehensive benchmark suite of video coding, we widely search available resources and include the most popular video codecs. VC-Bench is based on three high-quality codecs employing over 16 kernels. Several principles are listed below to make the benchmark suite representative, available, impartial and integrated.

- Popularity. Kernels included in VC-Bench must be extracted from applications which are widely used in modern video coding domain.
- Accessibility. The codecs must be under free license. This guarantees that we can get the source codes legally.

- Portability and fairness. Any machine-specific optimizations must be easily removed from the source codes, ensuring the benchmark suite easily run on different platforms.
- Integrality. Typical inputs, configurations and parameters must be provided for the usage of VC-Bench.

3.1 Codec Analysis

Peak Signal-to-Noise Ratio (PSNR) is commonly used as a metric for quality of lossy compression coding. We compare PSNR between codecs. We use default encoding configuration and input videos are listed in Table 1. The resolution of these typical videos ranges from 176x144 to 1920x1080.

Table 1 Video Sequences

Name	Resolution
Foreman	176x144
Mobile	352x288
Ducks_fly	1280x720
Ducks_fly	1920x1080

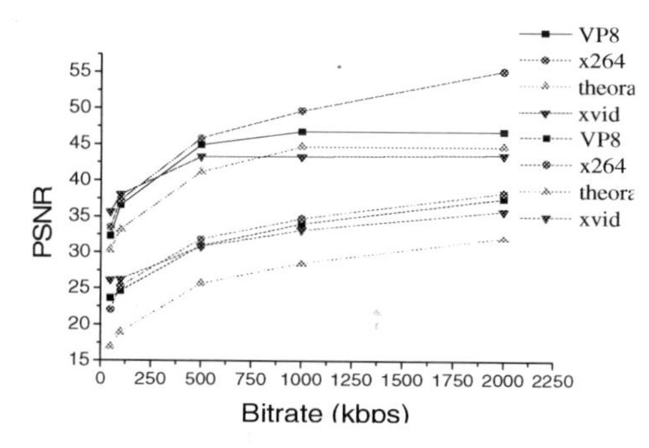

Fig. 3 PSNR analysis

Fig. 3 depicts PSNR of codecs and solid line represents 176x144 300 frames video while dotted line represents 352x288. All of X264, VP8 and XVID perform well with small differences in Fig. 3. Theora exhibits a lower encoding quality and this penalty rises along with the increase of resolution. X264 performs the best on both encoding quality and compression rate. When the resolution of video sequence is larger than 352x288, performance of X264, XVID and VP8 tends to get closer. This trend remains when the video resolution rises to 1920x1080. The coding efficiency of X264 and XVID are also superior [12]. Therefore, we choose X264, XVID and VP8 to build up VC-Bench. We exclude Theora due to its low coding quality; we use it only for PSNR comparison purpose.

3.1.1 X264

X264 is a free software library for encoding video streams into the H.264/MPEG-4 AVC format [9]. X264 implements a number of technologies to enhance coding quality and efficiency. The newest version of X264 implements Psychovisual Rate–distortion optimization which investigates the relationship between physical stimuli and sensations. It also uses Macroblock-tree rate control to adjust the quality by tracking how often MBs are used for predicting future frames. These methods make X264 more coding efficient and precise than other contemporary H.264 codecs.

X264 has five key phases as mentioned above. It provides five search algorithms in motion estimation (ME), such as hexagon search, and uses both Sum of Squares Differences (SSD) and Sum of Absolute Transformed Differences (SATD) as metrics to identify block matching. It also implements these search methods on subpixel and quarter pixel to enhance accuracy. Discrete Cosine Transform (DCT) is applied because of "energy compaction" property [13]. X264 implements adaptive quantization in two modes using Variance Adaptive Quantization (VAQ). VAQ is more time consuming but has higher compression rate than normal quantization. The output then goes into de-quantization process to construct reference frames. Each line of MBs within a frame is filtered by an adaptive loop filter to get rid of the blocking artifact on block boundaries. Finally, X264 applies Context-adaptive variable-length coding (CAVLC) and Context-adaptive binary arithmetic coding (CABAC) to encode the output of quantization into final data file which can be stored in disc or transmitted on the Internet. Therefore, the version we choose for VC-Bench is X264 0.116.x. We select DCT, inverse DCT, trellis quantization and loop filter.

3.1.2 VP8

VP8 is an open video compression format created by On2 Technologies. It is the eighth generation of VPX series. The implementation was released in 2010. VP8's license is free and the source code is available on its homepage [14]. Compared to X264 and XVID, VP8 includes both Discrete Cosine Transform (DCT) and Walsh-Hadamard Transformation (WHT) [15] which only act on 4x4 pixels. Moreover, during the quantization phase, VP8 provides three modes: fast, regular and strict. It also implements two specific quantization methods for luma and chroma respectively. Therefore, we select VP8 v1.0.0. VC-Bench includes both DCT and WHT. All of the quantization modes and a less smart loop filter, compared to loop filter of X264, are selected too.

3.1.3 XVID

XVID is a video codec library based on MPEG-4 Part 2 Advanced Simple Profile (ASP). It implements features such as b-frames, global and quarter pixel motion compensation, lumi masking, three quantization modes (trellis, H.263 and MPEG), and custom quantization matrices. XVID is an open source video codec under free license. It is also widely used in free multimedia players and transcoder applications. Unlike X264, the implementations of algorithm in XVID are different. For example, in quantization phase, XVID provides four modes. They are h263 quantization from standard H.263, mpeg quantization, normal quantization, and trellis quantization. Additionally, XVID only provides 8x8 pixels DCT transform while X264 provides both 8x8 and 4x4 DCT transform. Consequently, we use XVID-1.3.2 and select the forward DCT, inverse DCT transformation, and loop filter. VC-Bench also includes all the quantization modes.

3.2 *Hotspots Analysis*

We focus on computational kernels because these kernels help us to identify bottlenecks in coding process. Bottlenecks are usually caused by various aspects especially intensive computational operations and memory accesses. They are the most time consuming part. Thus, we do hotspots analysis to show the distribution of time clearly. We count the most time consuming functions in single-threaded versions of all the three selected codecs. Results help us to find computational kernels. A thorough understanding of these kernels can facilitate possible software and hardware optimizations. We choose foreman with resolution 176x144 as input video sequence. We disable the multithread mode and eliminate assembly optimizations. The detail of platform will be discussed later in Table 3.

3.2.1 Hotspots in X264

The computing intensive nature of X264 codec is well captured by Fig. 4a. This figure shows that the majority of running time is occupied by motion estimation (ME) and quantization. As bitrate increases, time occupation of ME declines while that of quantization rises. The two curves cross when the bitrate reaches 1000 kbps, and then quantization becomes the major time consuming part. When the bitrate is between 50 kbps and 100 kbps, ME accounts for nearly 50% of execution time in x264 while another 15% of time is spent on quantization. When the bitrate rises to 2000kbps, ME occupies 25% while quantization accounts for 45%. Although other three phases (Discrete Cosine Transform, Loop filter and Entropy Coding) have a smaller running time compared to ME and quantization, we also list them because they are indispensable phases of video coding process. They together occupy 10% of the program running time.

Furthermore, as a data intensive application, total execution time of codecs scales with the increase of input video resolution. This result is presented in Fig. 5. We fix the bitrate in 500kbps and use the four video sequences listed in Table 1.

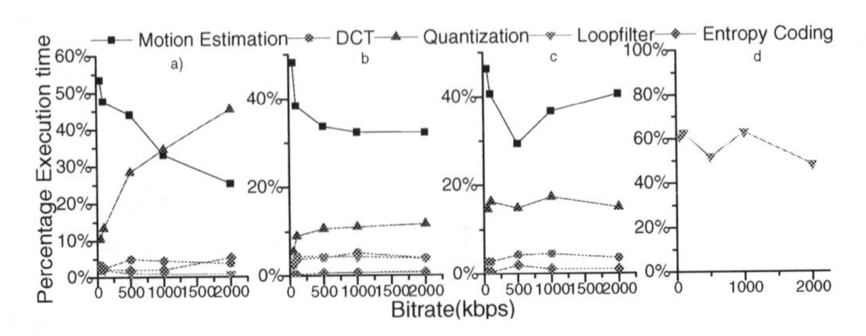

Fig. 4 Hotspots. a X264 b VP8 c&d XVID

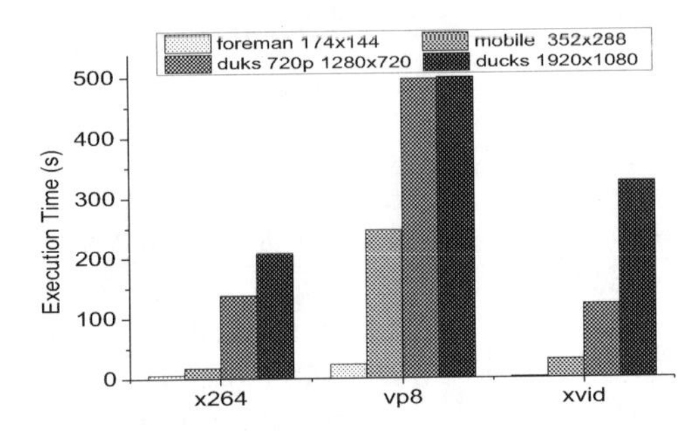

Fig. 5 Execution time

3.2.2 Hotspots in VP8

We generate VP8's hotspots profiling data in Fig. 4b. For a fixed bitrate, like 500 kbps, time consumption of ME is two times larger than that of Quantization. As higher bitrate puts less constraint on the precision of block matching, time occupancy of ME declines while that of quantization rises with the increase of bitrate. The difference between VP8 and X264 is obvious. Execution time of quantization in VP8 never reaches 20% and the curve is nearly horizontal after 500 kbps while X264's quantization rises to 45%. The reason is that X264 uses Trellis [16] quantization, which is much more complex and time consuming than normal quantization methods. Fig. 5 reflects that VP8's default configuration is more time consuming than other two codecs no matter what resolution of video is.

3.2.3 Hotspots in XVID

Fig. 4c and 4d summarize the time distribution of XVID. We use Fig. 4d to represent XVID loop filter because we derive it from decoding process which is separate from encoding process. Compared to VP8 and X264, the overall trend of time consumption remains unchanged. The ratio of ME drops down from 44% and reaches the lowest point 29% at 500 kbps. Then it slowly increases and finally constitutes 39%. The proportion of quantization fluctuates around 18%. DCT and Entropy coding together occupy less than 10%. In particular, Loop filter is the dominated part in XVID decoding and it consumes approximately 60% of decoding time.

3.2.4 Description of Hotspots

We notice that ME is the most time consuming part under default configuration because it operates iteratively on pixel granularity. This observation is portrayed by the scaling of execution time in Fig. 4. As we discussed above, the function of ME is block matching. With this method, the output residual of block subtraction will be smaller and final compression rate will be higher. However, this paper focuses on quantization, DCT and Loop Filter except ME for two reasons. First, there have been various new algorithms and optimizations focusing on Motion Estimation and all of them strive to decrease ME's time consumption [17, 18]. Second, the major operations of ME are subtraction and calculation of SSD as well as SATD. There are many computing instructions with less memory accesses and dependence. Therefore, parallelization is easy to implement so that ME's time consumption will reduce significantly [19, 20, 21]. Fig. 6 gives an example of SIMD acceleration. We present the top ten time consuming functions in X264 and result shows that SIMD can largely reduce the time of ME compared to Fig. 4. Quantization becomes the dominated part in Fig. 6.

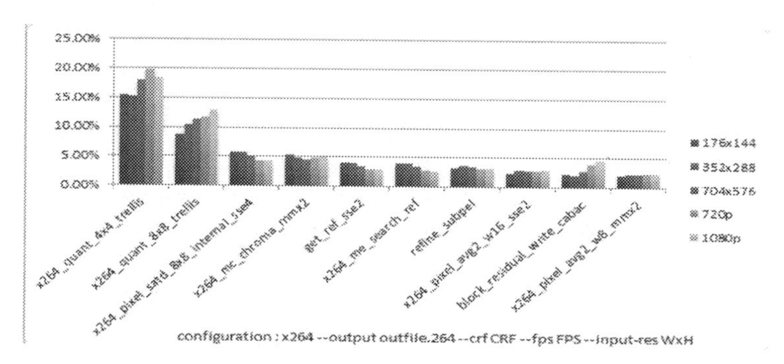

Fig. 6 Hotspots in X264

The significant decline in time consumption of ME makes a role exchange in hotspots. Phases like quantization and DCT become the major time comsumption as Fig. 6 shows. There are few analysis and optimizations on phases except ME. Knowing these phases' behavior will bring inspiration on optimizations. Therefore, we first choose quantization, DCT, and loop filter to construct VC-Bench in order to present their computing and memory access characteristics.

3.3 *Kernel Extraction*

Long running time and the shortage of performance tools make it hard to analyze the behavior of hotspots. Besides, video codecs contains various redundant codes, such as command parsing, video quality analyzing and so on. Removing these codes can largely reduce code size and leave computational kernels alone. It is better for performance analysis especially on embedded platforms. Therefore, we extract hotspots of video coding process and use these extracted fragments as test programs in VC-Bench. Each program is written in C language that can run independently on different platforms. We eliminate machine-specific optimizations. Table 2 lists the components of VC-Bench.

Table 2 VC-Bench

Benchmark	Description	Input data set	Characteristic
x264_dct_idct	DCT & IDCT of x264	S,N,L	Data intensive
vp8_dct_idct	DCT & IDCT of vp8	S,N,L	Data intensive
xvid_dct_idct	DCT & IDCT of xvid	S,N,L	Data intensive
x264_quant	CABAC and CAVLC Trellis quantization	S,N,L	Computation and data intensive
vp8_quant	Quantization of VP8(fast, regular and strict)	S,N,L	Data intensive
xvid_h263	Quantization in xvid use h263 method.	S,N,L	Data intensive
xvid_mpeg	Quantization in xvid use mpeg method.	S,N,L	Data intensive
xvid_trellis	Quantization in xvid use trellis.	S,N,L	Data intensive
x264_deblock	Loop filter of x264	S,N,L	Computation intensive
vp8_deblock	Loop filter of vp8	S,N,L	Computation intensive
xvid_deblock	Loop filter of xvid	S,N,L	Computation intensive

Three input data sets are provided from videos in Table 1. We add file writing instructions in original codecs to gather the input data for VC-Bench. The sets labeled S, N, and L are derived from video with resolution 176x144, 352x288 and 1280x720 respectively. All the input data are in text document format and we use default encoding configurations to gather the data sets.

We verify the correctness of extracted kernels from two aspects. First, we write the outputs of kernels back to codecs to complete subsequent coding operations. When finished, we compare its MD5 with original codec's output. This is functional verification. Second, we compare total execution time proportion between extracted kernels and integrated codec. We call this comparison timing verification. Fig. 7 depicts the basic structure of programs in VC-Bench and also shows the functional verification process.

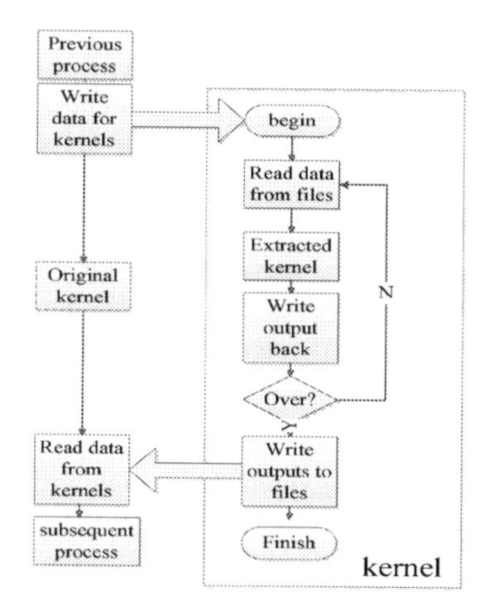

Fig. 7 Functional Verification

4 Experiment Result

4.1 Experiment Setup

We conduct our analysis on a server with two Intel Xeon E5520 processors with a three-level cache hierarchy. Both L1 and L2 are private to each core while L3 is shared among all cores. L1 cache further splits into 32KB Instruction Cache and 32KB Data Cache. Table 3 summarizes parameters of the platform. In order to collect architectural profiling data, we use Intel VTune Amplifier XE 2011 [22], a tool that provides an interface to the processor performance units. It provides kinds of code profiling including thread profiling, stack sampling and hardware event sampling. VTune can find specific tuning opportunities like branch divergence and cache performance. Note that all the subsequent results we present are filtered so that file accesses operations are not captured.

Table 3 Configuration

Feature	Description
OS	Linux 2.6.18-128.e15
Processor	Two Intel ® Xeon ® CPU E5520 2.27GHz
Memory Size	70GB
L1 I-cache	32KB, 4-way associative
L1 D-cache	32KB, 8-way associative
L2 Cache	256KB, 8-way associative
L3 Cache	8MB, 16-way associative

4.2 Computational Performance

We begin results analysis focusing on computational properties. We use small input data set. Fig. 8 shows timing verification. It portrays the proportion of cycles for single-threaded runs of VC-Bench versus integrate codecs and the results are normalized. The overall proportion among DCT, quantization and loop filter is approximately similar between extracted kernels and original codecs. This result indicates that our work is reasonable and can well characterize phases of original codecs with 90% less codes. We also verified that normal input set and large input set perform the similar results.

Fig. 9 clearly shows instructions per cycle (IPC) of each benchmark. All the columns fluctuate near 1.5 and the variance is small except for H263 quantization of XVID. This figure implies that the instruction level parallelism can be extracted from these benchmarks is low. Dependence of instructions and data is ubiquitous within majority of these benchmarks. Thus, instruction level parallelism may not efficiently drive performance improvement which leaves data level and thread level parallelism to enhance properties in video coding workloads [23].

Cycle accounting for total executed micro-operations is an effective technique to identify bottlenecks for performance tuning. One compiled operation, or micro-operation, contains four phases namely "issued", "dispatched", "executed" and "retired". The total cycles can be divided into those where micro-operations are dispatched to the execution units and those where no micro-operations are dispatched, which are thought of as execution stalls [24]. We graph the percentage of stalled cycles in Fig. 10. The average stalled cycle ratio is about 7% except for H263 and trellis quantization of XVID. XVID's H263 quantization exhibits an outstanding cycle stall around 24%. The stalled cycles are brought by operations such as memory accesses and the very long latency instructions like divide and square.

Branch mispredictions can also bring execution inefficiencies in processor. This penalty includes wasted work on executing incorrectly predicted path, cycle lost on flushing execution pipeline and time spending on waiting right operations reaching execution units. We present branch misprediction rate in Fig. 11 as well as total runtime branch instructions' proportion in Fig. 13. We note that XVID exhibits a higher misprediction than X264 and VP8 on average. XVID mpeg-quantization module exhibits about 18% and others are below 10%. The reason of high mpeg quantization is that its branch condition changes irregularly. Current history based branch predictors do not suit this kind of pattern.

Additionally, we examine instruction fetch stall rate and we find that this stall rate is very small with the exception of XVID loop filter. XVID loop filter's high stall rate is because of its high L1 instruction cache miss rate. The reason is discussed later.

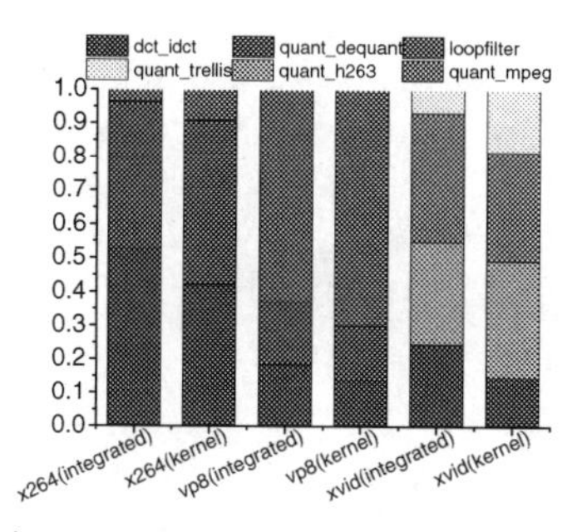

Fig. 8 Execution Cycles (kernel vs codec)

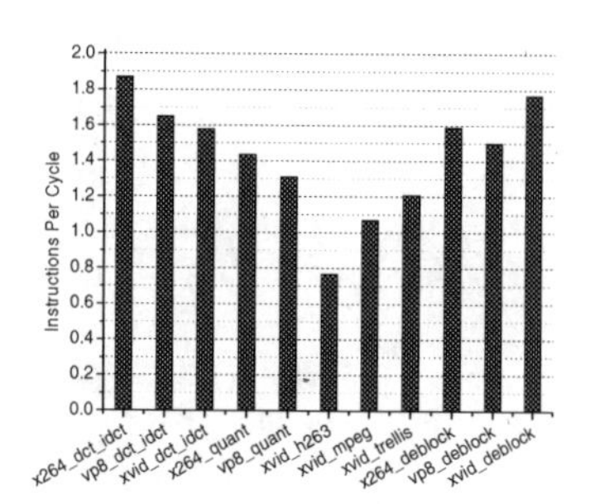

Fig. 9 IPC

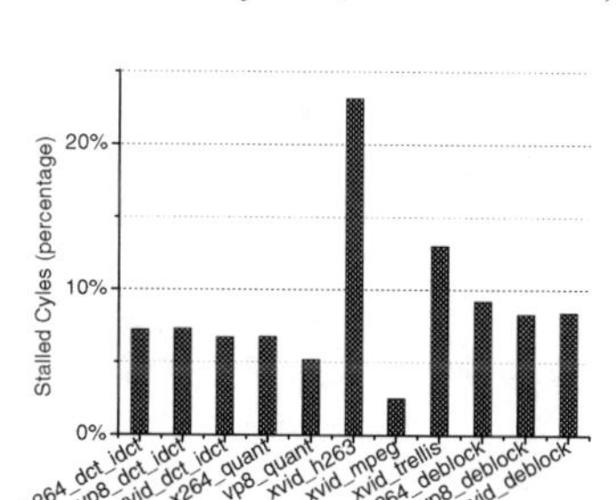

Fig. 10 Stalled Cycles

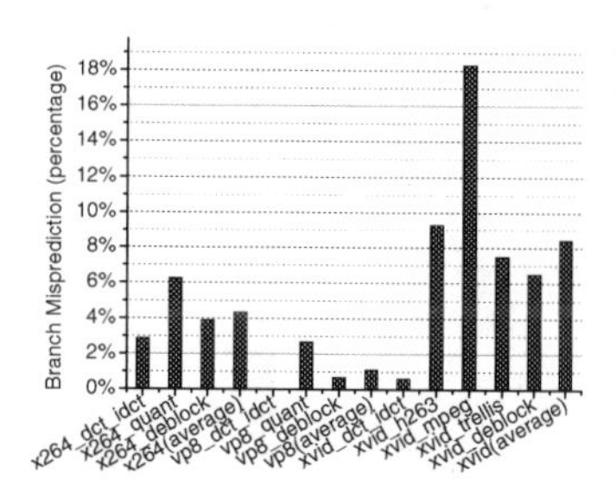

Fig. 11 Branch Misprediction

4.3 Memory Access Performance

Most of benchmarks in Table 2 are data intensive programs. Fig. 13 portrays the total runtime memory accesses and branch instructions. Long latency instructions and memory accesses are two primary reasons causing stalls of CPU. If memory accesses frequently miss in the cache, large amount of memory instructions certainly bring dominant penalty. However, we note that L1 D-cache hit rate is very high. It is over 99% among all the benchmarks. Two reasons may bring this result. First, DCT and quantization operate on small data blocks iteratively with regular memory access behaviors. These operations are generally complex and contain plenty long time computing instructions like square and division. Therefore, when a cache line is loaded, the following several loads and stores can hit the cache and the penalty brought by a small number of cache miss is negligible compared to computing time. Second, we profile the results with the hardware prefetching which may benefit the cache properties.

Fig. 12 plot L1 cache performance. For miss rates smaller than 0.01%, we use 0.01% to represent. The influence of hardware prefetching is also considered. In Fig. 12, L1 instruction catch miss is relatively low under 2.5% except XVID loop filter which exhibits over 30% instruction misses. This is because XVID uses macros instead of function calls while doing loop filter. As loop filter contains large number of loops, macro expansion after compiling makes object file nearly 340KB which is far beyond I-Cache capacity. When we use functions instead of macros, this miss rate largely reduces. Two benchmarks of VP8 show the lowest instruction cache miss rate partially because they contain less branches and the branch misprediction rate is lower as in Fig. 11. L1 data cache miss rate is also portrayed by Fig. 12 and all benchmarks' miss rate is under 0.5%. Most of them can benefit from prefetching even though the effect is not obvious.

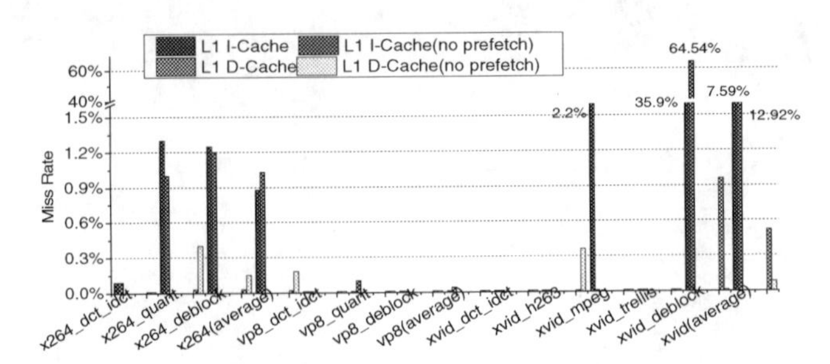

Fig. 12 L1 Cache performance

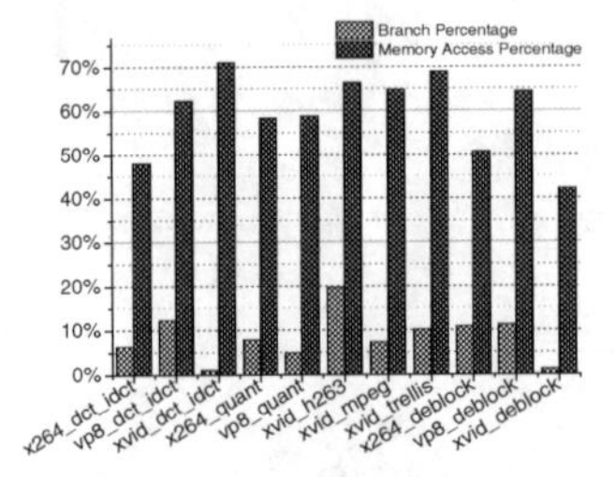

Fig. 13 Instruction distribution

Fig. 14 SIMD influence

4.4 *Implications*

Based on the performance results, we can derive some implications about possible architecture modifications. Adding SIMD instructions can reduce the cost of long latency instructions and improve overall performance [25]. For an example, we present SIMD influence on X264 DCT and IDCT kernel in Fig. 14. We normalize total execution cycles to 1 for comparison. We notice that SIMD can largely reduce execution cycles by 90% and enhance IPC up to 2.7. Branch misprediction and instruction starvation are also lowered by 34% and 26% respectively.

The major memory access latency is dominated by L1 latency due to extremely high L1 data cache hit rate. Large L2 cache and LLC consume processor chip area but do not yield a performance benefit. Taeho Kgil and his group showed that, for a particular class of throughput-oriented workloads, modern processors are extremely power inefficient, arguing that the chip area should be used for processing cores rather than caches [26]. Our results reinforce these findings, pointing out that the cache hierarchy needs to resize and reorganize targeting at video coding applications.

Finally, low instruction level parallelism (ILP) precludes effectively using full core width. This low ILP is caused by various instruction dependences which is an inherent characteristic of video coding applications. However, the nature of these applications makes them ideal to apply thread level parallelism (TLP). Optimizing computational kernels on accelerated architectures, such as GPU, will probably gain performance enhancement.

5 Conclusion

In this paper, we present a video coding benchmark suite (VC-Bench), which is composed of hotspots extracted from a wide range of video codecs. These hotspots cover three key phases of coding process namely discrete cosine transform (DCT), quantization and loop filter. We also provide a spectrum of inputs to help understanding the influence of input size on performance. Experimental results confirm the reasonability of methodology through both functional and timing verification. VC-Bench can well represent the characteristics of integrate video codecs with 90% less in code size and overcome the shortages of performance analyzing tool. Measured hardware results show high L1 cache hit rate which imply that deep cache hierarchy is not necessary.

In future work, we intend to incorporate the other two phases (Motion Estimation and entropy coding). To further explore the parallelism characteristics of processors on video coding, we also plan to add multi-threaded version of computational kernels to VC-Bench. Besides, an input data generator will be developed to ease the transfer of VC-Bench.

Acknowledgment. This work is supported financially by the National Basic Research Program of China under contract 2011CB302501, the National Hi-tech Research and Development Program of China under contracts 2012AA010902 and 2012AA010901.

References

1. Wiegand, T., Sullivan, G.J., Bjøntegaard, G., Luthra, A.: Overview of the H.264/AVC video coding standard. IEEE Trans. Circuits Syst. Video Technol. 13(7), 560–576 (2003)
2. Bankoski, J., Wilkins, P., Xu, Y.: VP8 Data Format and Decoding Guide (January 2011)
3. Clemons, J., Zhu, H., Savarese, S., Austin, T.: MEVBench: A Mobile Computer Vision Benchmarking Suite. In: 2011 IEEE International Symposium Published on Workload Characterization (IISWC), November 6-8 (2011)
4. Venkata, S.K., Ahn, I., Jeon, D., Gupta, A., Louie, C., Garcia, S., Belongie, S., Taylor, M.B.: SD-VBS: The San Diego Vision Benchmark Suite. In: IEEE International Symposium Published on Workload Characterization, IISWC 2009, October 4-6 (2009)

5. Lee, C., Potkonjak, M., Mangione-Smith, W.H.: MediaBeinch: A Tooi for Evaluating and Synthesizing Multimedia and Communications Systems. In: Published on MICRO 30 Proceedings of the 30th annual ACM/IEEE international symposium on Microarchitecture (1997)
6. Fritts, J.E., Steiling, F.W., Tucek, J.A., Wolf, W.: MediaBench II Video: Expediting the next generation of video systems research. In: Embedded Processors for Multimedia and Communications II, pp. 79–93 (2005)
7. H.264/14496-10 AVC Reference Software Manual (January 2009)
8. FFmpeg Multimedia System (2005), http://ffmpeg.mplayerhq.hu/
9. X264. A Free H.264/AVC Encoder (2006), http://developers.videolan.org/x264.html
10. Home of the xvid codec, http://www.xvid.org/
11. Alvarez, M., Salami, E., Ramírez, A., Valero, M.: HD-VideoBench. A Benchmark for Evaluating High Definition Digital Video Applications. In: IEEE 10th International Symposium Workload Characterization, IISWC 2007, September 27-29 (2007)
12. Sixth MPEG-4 AVC/H.264 Video Codecs Comparison - Short Version, http://compression.ru/video/codec_comparison/h264_2010/#Video_Codecs
13. Rao, K.R., Yip, P.: Discrete Cosine Transform: Algorithms, Advantages, Applications, Boston, MA (1990)
14. WebM, an open web media project (2010), http://www.webmproject.org/tools/vp8-sdk/
15. Partt, W.K., Kane, J., Andrews, H.C.: Hadamard transform image coding (1969)
16. Merritt, L.: Notes on the implementation of trellis quantization in H.264 (November 2005)
17. Zhu, S., Ma, K.-K.: A New Diamond Search Algorithm for Fast Block Matching Motion Estimation. IEEE Transactions Image Processing (2000)
18. Lin, W., Panusopone, K., Baylon, D., Sun, M.-T.: A New Class-based Early Termination Method for Fast Motion Estimation in Video Coding. In: IEEE International Symposium Circuits and Systems, ISCAS 2009, May 24-27 (2009)
19. Ko, Y., Yi, Y., Ha, S.: An efficient parallel motion estimation algorithm and X264 parallelization in CUDA. In: Sch. of EECS, Seoul Nat. Univ., Seoul, South Korea, November 2-4 (2011)
20. Wang, J., Lu, R., Qiu, C., Gao, P., Lu, Y., Yu, W.: FPGA-accelerated Design of Motion Estimation for H.264 HDTV. In: Second International Conference on Information and Computing Science, ICIC 2009, May 21-22 (2009)
21. Momcilovic, S., Sousa, L.: A Parallel Algorithm For Advanced Video Motion Estimation on Multicore Architectures. In: International Conference on Complex, Intelligent and Software Intensive Systems, CISIS 2008, March 4-7 (2008)
22. Intel® VTune™ Amplifier XE (2011), http://software.intel.com/en-us/articles/intel-vtune-amplifier-xe/
23. Jacobs, T.R., Chouliaras, V.A., Mulvaney, D.J.: Thread-parallel MPEG-2, MPEG-4 and H.264 video encoders for SoC multi-processor architectures. IEEE Transactions on Consumer Electronics 52(1) (February 2006)
24. Intel® 64 and IA-32 Architectures Optimization Reference Manual (April 2011)
25. Lee, J., Moon, S., Sung, W.: H.264 decoder optimization exploiting SIMD instructions. School of Electrical Engineering, Seoul National University (2004)
26. Kgil, T., D'Souza, S., Saidi, A., et al.: PicoServer: using 3D stacking technology to enable a compact energy efficient chip multiprocessor. In: ASPLOGS (2006)

Associative Models for Encrypting Monochromatic Images

Elena Acevedo, Antonio Acevedo, Fabiola Martínez, and Ángel Martínez

Abstract. This work describes a novel method for encrypting monochromatic images using an associative approach. The image is divided in blocks which are used to build *max* and *min* Alpha-Beta associative memories. The key is private and it depends on the number of blocks. The main advantage of this method is that the encrypted image does not have the same size than the original image; therefore, since the beginning the adversary cannot know what the image means.

Keywords: Artificial Intelligence, Associative approach, Encryption, monochromatic images.

1 Introduction

The advent of personal computers and the Internet has made it possible for anyone to distribute worldwide digital information. However, there are many applications that need to protect their information from people who can steal important data. Therefore, it is important to apply a method for hiding information, i.e., it is necessary to encrypt data.

Traditional image encryption algorithms are private key encryption standards (DES and AES), public key standards such as Rivest Shamir Adleman (RSA), and the family of elliptic-curve-based encryption (ECC), as well as the international data encryption algorithm (IDEA).

Current encryption algorithms can be classified into different techniques such as optical [1-6], value transformation [7-11], pixels position permutation [12-14] and chaos-based [15-19].

Elena Acevedo · Antonio Acevedo · Fabiola Martínez · Ángel Martínez
Escuela Superior de Ingeniería Mecánica y Eléctrica Zacatenco, Instituto Politécnico Nacional, Mexico City, Mexico
e-mail: {eacevedo,macevedo,fmartinezzu}@ipn.mx, josekun13@gmail.com

R. Lee (Ed.): *SNPD*, SCI 492, pp. 117–128.
DOI: 10.1007/978-3-319-00738-0_9

In this paper we propose a novel encryption method for monochromatic images by using an associative approach. In particular, we use the Alpha-Beta associative memory which has demonstrated to be a suitable tool for the Pattern Recognition area.

2 Associative Models

Associative models simulate the behavior for learning of the human brain. These models associate stimuli with responses, once the algorithm has associated all stimuli with their corresponding responses, the Associative Memory is able to recall a response when a stimulus is presented to it. These stimuli and responses are patterns represented by vectors. The task of association is called Training Phase and the Recognizing Phase allows recovering patterns. This process is illustrated in Figure 1. The stimuli are the input patterns represented by the set $\mathbf{x} = \{x^1, x^2, x^3, \ldots, x^p\}$ where p is the number of associated patterns. The responses are the output patterns and are represented by $\mathbf{y} = \{y^1, y^2, y^3, \ldots, y^p\}$. Representation of each vector x^μ is $x^\mu = \{x_1^\mu, x_2^\mu, \ldots, x_n^\mu\}$ where n is the cardinality of x^μ. The cardinality of vectors y^μ is m, then $y^\mu = \{y_1^\mu, y_2^\mu, \ldots, y_m^\mu\}$.

Training Phase

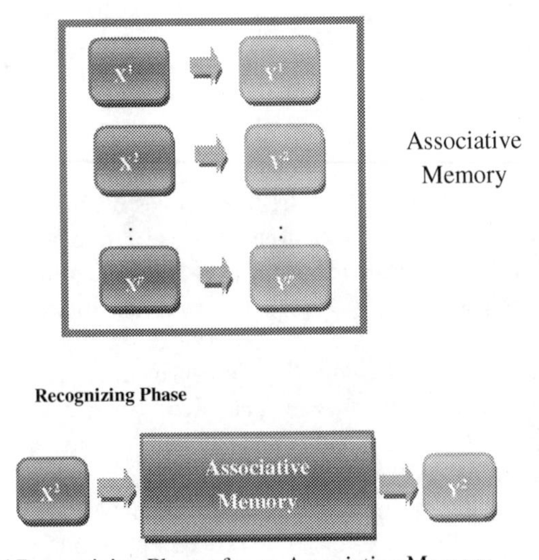

Fig. 1 Training and Recognizing Phases for an Associative Memory

From Figure 1, the set of associations is called the fundamental set and is represented as follows:

$$\{(\mathbf{x}^\mu, \mathbf{y}^\mu) \mid \mu = 1, 2, \ldots, p\}$$

A memory is ***Autoassociative*** if it holds that $x^{\mu} = y^{\mu}$ $\forall \mu \in \{1, 2, ..., p\}$, then one of the requisites is that $n = m$.

A memory is ***Heteroassociative*** when $\exists \mu \in \{1, 2, ..., p\}$ for which $x^{\mu} \neq y^{\mu}$. Notice that there can be heteroassociative memories with $n = m$.

Now, we will describe the Alpha-Beta associative model [20].

Alpha-Beta Associative Memories inherit the name from the operators α and β. The operator α is applied in the training phase to associate the pairs of patterns belonging to the fundamental set. In the recognizing phase, the β operator is applied to recall the corresponding output pattern when an input pattern is presented to the associative memory.

The sets $A = \{0, 1\}$ and $B = \{0, 1, 2\}$, the α and β operators, along with the usual $\wedge$ (minimum) and $\vee$ (maximum) operators form the algebraic system $(A, B, \alpha, \beta, \vee, \wedge)$ which is the mathematical basis for the Alpha-Beta associative memories. The operator α is defined in tabular form as Table 1 shows.

Table 1 Definition of operator α: $A \times A \to B$

x	Y	α (x, y)
0	0	1
0	1	0
1	0	2
1	1	1

The β operator is defined in Table 2.

Table 2 Definition of operator β: $B \times A \to A$

x	y	β(x, y)
0	0	0
0	1	0
1	0	0
1	1	1
2	0	1
2	1	1

We have two types of Alpha-Beta heteroassociative memories: *max* (**V**) and *min* (**Λ**). The operator $\otimes$ is used to build both types of memories as Equation (1) shows:

$$\left[\mathbf{y}^{\mu} \otimes \left(\mathbf{x}^{\mu}\right)^{t}\right]_{ij} = \alpha\left(y_i^{\mu}, x_j^{\mu}\right) \tag{1}$$

where $\mu \in \{1, 2,..., p\}$, $i \in \{1, 2,..., m\}$ and $j \in \{1, 2,..., n\}$

Training phase

Step 1. For each $\mu = 1, 2, ..., p$ and from the pair $(\mathbf{x}^\mu, \mathbf{y}^\mu)$ we use Equation (1) to build

$$\left[\mathbf{y}^\mu \otimes (\mathbf{x}^\mu)^t\right]_{mxn}$$

Step 2. The maximum ($\vee$) and the minimum ($\wedge$) operators are applied to the matrices obtained in step 1, Equations (2) and (3) illustrate this process.

$$\mathbf{V} = \bigvee_{\mu=1}^{p} [\mathbf{y}^\mu \otimes (\mathbf{x}^\mu)^t] \tag{2}$$

$$\mathbf{\Lambda} = \bigwedge_{\mu=1}^{p} [\mathbf{y}^\mu \otimes (\mathbf{x}^\mu)^t] \tag{3}$$

The ij-th entries of both matrices are given by Equations (4) and (5).

$$v_{ij} = \bigvee_{\mu=1}^{p} \alpha(y_i^\mu, x_j^\mu) \tag{4}$$

$$\lambda_{ij} = \bigwedge_{\mu=1}^{p} \alpha(y_i^\mu, x_j^\mu) \tag{5}$$

Recognizing phase

The pattern x^ω is presented to the Alpha-Beta heteroassociative memories $\mathbf{V}$ and $\mathbf{\Lambda}$ and Equations (6) and (7) are applied.

$$\left(\mathbf{V}\Delta_\beta \mathbf{x}^\omega\right)_i = \bigwedge_{j=1}^{n} \beta\left(v_{ij}, x_j^\mu\right) \tag{6}$$

$$\left(\mathbf{\Lambda}\nabla_\beta \mathbf{x}^\omega\right)_i = \bigvee_{j=1}^{n} \beta\left(\lambda_{ij}, x_j^\mu\right) \tag{7}$$

Due to the dimensions of the matrices $\mathbf{V}$ and $\mathbf{\Lambda}$ are m x n and x^ω is a column vector with dimension n, the results from the latter operation must be a column vector with dimension m whose i-th component is obtained by Equations (8) and (9).

$$\left(\mathbf{V}\Delta_\beta \mathbf{x}^\omega\right)_i = \bigwedge_{j=1}^{n} \beta\left(v_{ij}, x_j^\mu\right) \tag{8}$$

$$\left(\mathbf{\Lambda}\nabla_\beta \mathbf{x}^\omega\right)_i = \bigvee_{j=1}^{n} \beta\left(\lambda_{ij}, x_j^\mu\right) \tag{9}$$

3 Alpha-Beta Encryption

The algorithm for encrypting monochromatic images using Alpha-Beta associative memories is described in this section. An illustrative example is presented for this purpose.

We will apply our proposal to the image showed in Figure 2.

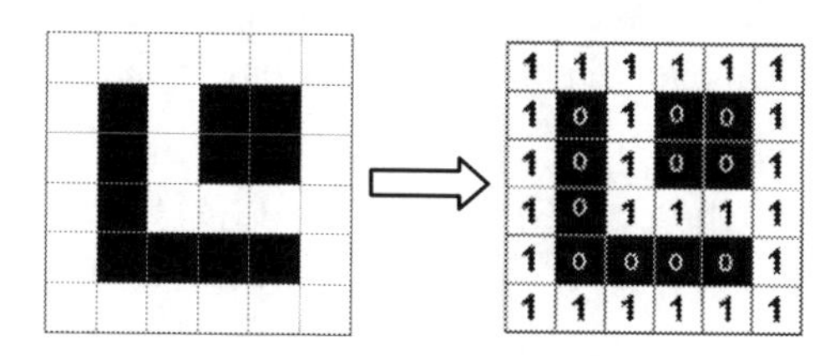

Fig. 2 Monochromatic image used for illustrating our proposal, white color is represented by 1 and black color is represented by 0

The image is converted to a binary matrix where the white color is represented by the number 1 and a number 0 is used to represent the black color. For this example, the binary matrix is divided in four parts. The image can be divided in *a* rows and *b* columns as the user decides. Now, we have four submatrices: y^1, y^2, y^3 and y^4, as Figure 3 shows.

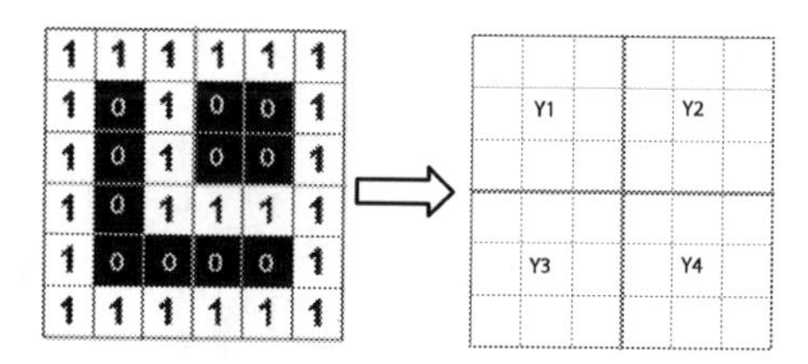

Fig. 3 The matrix is divided in 2 rows by 2 columns, and then we have four submatrices

Now, these submatrices are vectorized as follows:

$$y^1 = [1\ 1\ 1\ 1\ 0\ 1\ 1\ 0\ 1]$$
$$y^2 = [1\ 1\ 1\ 0\ 0\ 1\ 0\ 0\ 1]$$
$$y^3 = [1\ 0\ 1\ 1\ 0\ 0\ 1\ 1\ 1]$$
$$y^4 = [1\ 1\ 1\ 0\ 0\ 1\ 1\ 1\ 1]$$

These four vectors represent the output patterns of the associative memory. Each vector has 9 elements, i.e., $m = 9$.

The next step is to build the input patters which will be the private key for encryption. The number of elements of these vectors is the number of parts that the image was divided in, then the cardinality of patterns x is $n = a$ x b, in this example is 2 x 2 = 4. The four input patterns are *one-hot* vectors [21] and take the form:

$$x^1 = [1\ 0\ 0\ 0]$$
$$x^2 = [0\ 1\ 0\ 0]$$
$$x^3 = [0\ 0\ 1\ 0]$$
$$x^4 = [0\ 0\ 0\ 1]$$

These set of vectors, x and y, represent the Fundamental Set.

Equation (1) and the definition of α operator from Table 1 are applied to associate the first pair (x^1, y^1).

$$y^1 \otimes (x^1)^t = \begin{bmatrix} \alpha(1,1) & \alpha(1,0) & \alpha(1,0) & \alpha(1,0) \\ \alpha(1,1) & \alpha(1,0) & \alpha(1,0) & \alpha(1,0) \\ \alpha(1,1) & \alpha(1,0) & \alpha(1,0) & \alpha(1,0) \\ \alpha(1,1) & \alpha(1,0) & \alpha(1,0) & \alpha(1,0) \\ \alpha(0,1) & \alpha(0,0) & \alpha(0,0) & \alpha(0,0) \\ \alpha(1,1) & \alpha(1,0) & \alpha(1,0) & \alpha(1,0) \\ \alpha(1,1) & \alpha(1,0) & \alpha(1,0) & \alpha(1,0) \\ \alpha(0,1) & \alpha(0,0) & \alpha(0,0) & \alpha(0,0) \\ \alpha(1,1) & \alpha(1,0) & \alpha(1,0) & \alpha(1,0) \end{bmatrix} = \begin{bmatrix} 1 & 2 & 2 & 2 \\ 1 & 2 & 2 & 2 \\ 1 & 2 & 2 & 2 \\ 1 & 2 & 2 & 2 \\ 0 & 1 & 1 & 1 \\ 1 & 2 & 2 & 2 \\ 1 & 2 & 2 & 2 \\ 0 & 1 & 1 & 1 \\ 1 & 1 & 1 & 1 \end{bmatrix}$$

The same process is applied to the remaining pairs of vectors. The next step is to use Equations (2) and (3) to build *max* and *min* memories, and the results are:

$$V = \begin{bmatrix} 2 & 2 & 2 & 2 \\ 2 & 2 & 2 & 2 \\ 2 & 2 & 2 & 2 \\ 2 & 2 & 2 & 2 \\ 1 & 1 & 1 & 1 \\ 2 & 2 & 2 & 2 \\ 2 & 2 & 2 & 2 \\ 2 & 2 & 2 & 2 \\ 2 & 2 & 2 & 2 \end{bmatrix} \qquad \Lambda = \begin{bmatrix} 1 & 1 & 1 & 1 \\ 1 & 1 & 0 & 1 \\ 1 & 1 & 1 & 1 \\ 1 & 0 & 1 & 0 \\ 0 & 0 & 0 & 0 \\ 1 & 1 & 0 & 1 \\ 1 & 0 & 1 & 1 \\ 0 & 0 & 1 & 1 \\ 1 & 1 & 1 & 1 \end{bmatrix}$$

The binary matrices are converted to images. Both matrices V and Λ represent the encryption of the monochromatic image.

We can observe that the resulting encrypted images do not have the same dimensions than the original image. The latter has 6 x 6 pixels while the encrypted images have 4 x 9 pixels. This feature is the main advantage of our proposal because it is expected that the encrypted object has the same size than the original object. Therefore, since the beginning the Adversary cannot know what the image means.

The process to obtain the *ciphertext* is described as follows.

The information we need for decrypting the image is the number of rows and columns that the original image was divided in. In this case, a = 2 and b = 2, then the private key will have a cardinality of 2 x 2 = 4; thereby, the number of input vectors will also be 4. With this information, we build the *one-hot* vectors and they are presented to both associative memories. First, we apply Equation (8) to obtain output pattern y^1 with the *max* memory.

$$y^1 = V\Delta_\beta x^1 = \begin{bmatrix} 2 & 2 & 2 & 2 \\ 2 & 2 & 2 & 2 \\ 2 & 2 & 2 & 2 \\ 2 & 2 & 2 & 2 \\ 1 & 1 & 1 & 1 \\ 2 & 2 & 2 & 2 \\ 2 & 2 & 2 & 2 \\ 2 & 2 & 2 & 2 \\ 2 & 2 & 2 & 2 \end{bmatrix} \Delta_\beta \begin{bmatrix} 1 \\ 0 \\ 0 \\ 0 \end{bmatrix} = \begin{bmatrix} \beta(2,1) & \wedge & \beta(2,1) & \wedge & \beta(2,1) & \wedge & \beta(2,1) \\ \beta(2,1) & \wedge & \beta(2,1) & \wedge & \beta(2,1) & \wedge & \beta(2,1) \\ \beta(2,1) & \wedge & \beta(2,1) & \wedge & \beta(2,1) & \wedge & \beta(2,1) \\ \beta(2,1) & \wedge & \beta(2,1) & \wedge & \beta(2,1) & \wedge & \beta(2,1) \\ \beta(1,0) & \wedge & \beta(1,0) & \wedge & \beta(1,0) & \wedge & \beta(1,0) \\ \beta(2,1) & \wedge & \beta(2,1) & \wedge & \beta(2,1) & \wedge & \beta(2,1) \\ \beta(2,1) & \wedge & \beta(2,1) & \wedge & \beta(2,1) & \wedge & \beta(2,1) \\ \beta(2,0) & \wedge & \beta(2,0) & \wedge & \beta(2,0) & \wedge & \beta(2,0) \\ \beta(2,1) & \wedge & \beta(2,1) & \wedge & \beta(2,1) & \wedge & \beta(2,1) \end{bmatrix}$$

$$y^1 = V\Delta_\beta x^1 = \begin{bmatrix} 1 & \wedge & 1 & \wedge & 1 & \wedge & 1 \\ 1 & \wedge & 1 & \wedge & 1 & \wedge & 1 \\ 1 & \wedge & 1 & \wedge & 1 & \wedge & 1 \\ 1 & \wedge & 1 & \wedge & 1 & \wedge & 1 \\ 0 & \wedge & 0 & \wedge & 0 & \wedge & 0 \\ 1 & \wedge & 1 & \wedge & 1 & \wedge & 1 \\ 1 & \wedge & 1 & \wedge & 1 & \wedge & 1 \\ 1 & \wedge & 1 & \wedge & 1 & \wedge & 1 \\ 1 & \wedge & 1 & \wedge & 1 & \wedge & 1 \end{bmatrix} = \begin{bmatrix} 1 \\ 1 \\ 1 \\ 1 \\ 0 \\ 1 \\ 1 \\ 1 \\ 1 \end{bmatrix}$$

Now, we use Equation (9) to obtain y^1 with *min* memory

$$y^1 = \Lambda\nabla_\beta x^1 = \begin{bmatrix} 1 & 1 & 1 & 1 \\ 1 & 1 & 0 & 1 \\ 1 & 1 & 1 & 1 \\ 1 & 0 & 1 & 0 \\ 0 & 0 & 0 & 0 \\ 1 & 1 & 0 & 1 \\ 1 & 0 & 1 & 1 \\ 0 & 0 & 1 & 1 \\ 1 & 1 & 1 & 1 \end{bmatrix} \nabla_\beta \begin{bmatrix} 1 \\ 0 \\ 0 \\ 0 \end{bmatrix} = \begin{bmatrix} \beta(1,1) & \vee & \beta(1,1) & \vee & \beta(1,1) & \vee & \beta(1,1) \\ \beta(1,1) & \vee & \beta(1,1) & \vee & \beta(1,1) & \vee & \beta(1,1) \\ \beta(1,1) & \vee & \beta(1,1) & \vee & \beta(1,1) & \vee & \beta(1,1) \\ \beta(1,1) & \vee & \beta(1,1) & \vee & \beta(1,1) & \vee & \beta(1,1) \\ \beta(0,0) & \vee & \beta(0,0) & \vee & \beta(0,0) & \vee & \beta(0,0) \\ \beta(1,1) & \vee & \beta(1,1) & \vee & \beta(1,1) & \vee & \beta(1,1) \\ \beta(1,1) & \vee & \beta(1,1) & \vee & \beta(1,1) & \vee & \beta(1,1) \\ \beta(0,0) & \vee & \beta(0,0) & \vee & \beta(0,0) & \vee & \beta(0,0) \\ \beta(1,1) & \vee & \beta(1,1) & \vee & \beta(1,1) & \vee & \beta(1,1) \end{bmatrix}$$

$$y^1 = \Lambda\nabla_\beta x^1 = \begin{bmatrix} 1 & \vee & 1 & \vee & 1 & \vee & 1 \\ 1 & \vee & 1 & \vee & 0 & \vee & 1 \\ 1 & \vee & 1 & \vee & 1 & \vee & 1 \\ 1 & \vee & 0 & \vee & 1 & \vee & 0 \\ 0 & \vee & 0 & \vee & 0 & \vee & 0 \\ 1 & \vee & 1 & \vee & 0 & \vee & 1 \\ 1 & \vee & 0 & \vee & 1 & \vee & 1 \\ 0 & \vee & 0 & \vee & 0 & \vee & 0 \\ 1 & \vee & 1 & \vee & 1 & \vee & 1 \end{bmatrix} = \begin{bmatrix} 1 \\ 1 \\ 1 \\ 1 \\ 0 \\ 1 \\ 1 \\ 0 \\ 1 \end{bmatrix}$$

The same process is performed for the remaining input patterns (private keys) and the results for *max* memory are:

$$y^1 = \begin{bmatrix}1\\1\\1\\1\\0\\1\\1\\1\\1\end{bmatrix} \quad y^2 = \begin{bmatrix}1\\1\\1\\1\\0\\1\\1\\1\\1\end{bmatrix} \quad y^3 = \begin{bmatrix}1\\1\\1\\1\\0\\1\\1\\1\\1\end{bmatrix} \quad y^4 = \begin{bmatrix}1\\1\\1\\1\\0\\1\\1\\1\\1\end{bmatrix}$$

The obtained vectors with *min* memory are

$$y^1 = \begin{bmatrix}1\\1\\1\\1\\0\\1\\1\\0\\1\end{bmatrix} \quad y^2 = \begin{bmatrix}1\\1\\1\\0\\0\\1\\0\\0\\1\end{bmatrix} \quad y^3 = \begin{bmatrix}1\\0\\1\\1\\0\\0\\1\\1\\1\end{bmatrix} \quad y^4 = \begin{bmatrix}1\\1\\1\\0\\0\\1\\1\\1\\1\end{bmatrix}$$

Figure 4 shows the reconstruction of the images from the output vectors.

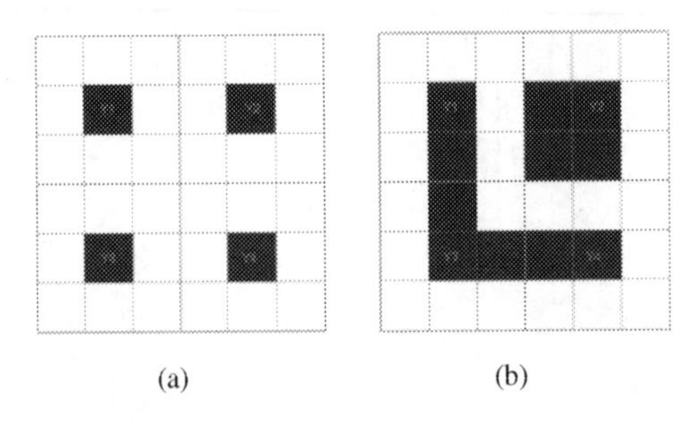

Fig. 4 The reconstructed images from Alpha-Beta *max* (a) and *min* (b) associative memories

In Figure 4(a) we can observe the recovered image with Alpha-Beta *max* associative memory and Figure 4(b) the shows the result from *min* memory. The original image is recovered with the *min* memory while *max* memory recovers just part of the original image.

4 Experiments and Results

We developed software to implement our proposal; the programming language was Visual C# 2010.

For testing the encryption method we used ten monochromatic images which are showed in Figure 5.

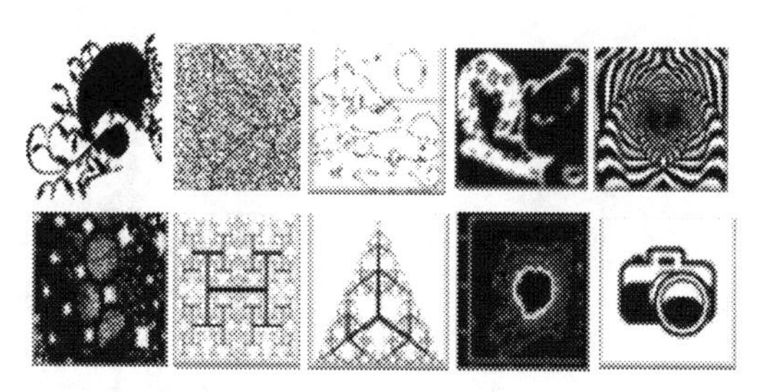

Fig. 5 Ten images for testing the performance of our proposal

Table 3 indicates the size in pixels of the ten images showed in Figure 5. The images can be square or not and the size can be variable.

Table 3 Size of pixels of the ten images

	Size in pixels
1	128 x 128
2	361 x 359
3	300 x 300
4	128 x 128
5	128 x 128
6	512 x 512
7	372 x 512
8	391 x 450
9	400 x 400
10	512 x 512

Table 4 shows the results of our proposal. We divided the ten images in different number of blocks which are indicated in the column 1 of the Table 4 aside the name of the image, the first number indicates the rows and the second indicates the columns. We applied Correlation and MSE to compare the original and recovered images. These measurements were applied to the results from *max* and *min* Alpha-Beta associative memories.

From Table 4, we can observe that *max* memory recovers the original image just when this image is divided in few blocks. On the other hand, *min* memory always recovers the original image.

Therefore, the Alpha-Beta associative memory which must be applied for encrypting monochromatic images is *min* memory.

Table 4 Results of recovering from the encryption method using *max* and *min* memories

Image	Max memory		Min memory	
	Correlation	MSE (%)	Correlation	MSE (%)
Im1 (3X3)	∞	35,1095	1	0
Im1 (4X3)	∞	34,1095	1	0
Im1 (5X5)	∞	35,4432	1	0
Im2 (1X2)	1	0	1	0
Im2 (5X7)	∞	37,9461	1	0
Im2 (10X15)	∞	37,9190	1	0
Im3 (2X2)	0,1064	6,0133	0,9999	0
Im3 (5X5)	∞	6,0867	0,9999	0
Im3 (12X9)	∞	6,1358	0,9999	0
Im4 (2X2)	0,7140	15,2465	1	0
Im4 (5X5)	0,7352	13,8183	1	0
Im4 (7X5)	∞	71,1746	1	0
Im5 (1X7)	0,5681	27,5062	1	0
Im5 (1X1)	1	0	1	0
Im5 (10X15)	∞	65,9165	1	0
Im6 (3X2)	0,5553	26,6547	1	0
Im6 (8X9)	∞	75,9347	1	0
Im6 (20X20)	∞	75,7036	1	0
Im7 (2X2)	0,4789	16,4661	1	0
Im7 (7X8)	∞	22,8784	1	0
Im7 (10X15)	∞	23,0302	1	0
Im8 (5X1)	0,2173	12,2820	1	0
Im8 (12X5)	∞	13,1776	1	0
Im8 (15X25)	∞	12,9817	1	0
Im9 (2X2)	0,8979	1,5931	1	0
Im9 (5X5)	0,2852	41,8318	1	0
Im9 (8X8)	0,0856	84,1825	1	0
Im10 (1X13)	∞	20,9810	0,9999	0
Im10 (5X25)	∞	21,3572	1	0
Im10 (8X15)	∞	20,8566	0,9999	0

5 Conclusions

Associative models have been applied in tasks such as pattern recognition and classification and they have shown competitive results with other approaches.

In this work, we demonstrated that the associative approach can be also applied to cryptography. In particular, Alpha-Beta associative memory *min*-type proved to be a suitable tool for encrypting monochromatic images.

The main advantage of this encryption algorithm is that the encrypted image does not have the same dimensions than the original image; therefore, it is more difficult to guess the meaning of the encryption.

The limitation of our proposal is that the encrypted image can be just stored or transmitted in a noise-free media. If any bit is changed then the original image cannot be recovered. However, this is the first time that Alpha-Beta associative memories are used to cryptography with positive results.

Acknowledgments. The authors would like to thank the Instituto Politécnico Nacional (COFAA, SIP and EDI), and SNI for their economic support to develop this work.

References

1. Tao, R., Xin, Y., Wang, Y.: Double image encryption based on random phase encoding in fractional Fourier domain. Opt. Express 15, 16067–16079 (2007)
2. Ge, F., Chen, L., Zhao, D.: A half-blind image hiding and encryption method in fractional Fourier domains. Opt. Commun. 281(17), 4254–4260 (2008)
3. Liu, Z., Li, Q., Dai, J., Sun, A., Liu, S., Ahmad, M.: A new kind of double encryption by using a cutting spectrum in the 1-D fractional Fourier transform domains. Opt. Commun. 282(8), 1536–1540 (2009)
4. Wang, B., Zhang, Y.: Double images hiding based on optical interference. Opt. Commun. 282(8), 3439–3443 (2009)
5. Meng, X., Cai, L., Wang, Y., Yang, X., Xu, X., Dong, G., et al.: Digital image synthesis and multiple-image encryption based on parameter multiplexing and phase-shifting interferometry. Opt. Lasers Eng. 47, 96–102 (2009)
6. Weng, D., Zhu, N., Wang, Y., Xie, J., Liu, J.: Experimental verification of optical image encryption based on interference. Optics Communications 284(10), 2485–2487 (2011)
7. Chen, R.J., Lai, J.L.: Image security system using recursive cellular automata substitution. Pattern Recognition 40, 1621–1631 (2007)
8. Guo, Q., Liu, Z., Liu, S.: Color image encryption by using Arnold and discrete fractional random transforms in HIS space. Optics and Lasers in Engineering 48(12), 1174–1181 (2010)
9. Tao, R., Meng, X.Y., Wang, Y.: Image encryption with multiorders of fractional fourier transforms. IEEE Transactions on Information Forensics and Security 5(4), 734–738 (2010)
10. Liu, Z., Xu, L., Lin, C., Dai, J., Liu, S.: Image encryption scheme by using iterative random phase encoding in gyrator transform domains. Optics and Lasers in Engineering 49(4), 542–546 (2011)
11. Liu, Z., Gong, M., Dou, Y., et al.: Double image encryption by using Arnold transform and discrete fractional angular transform. Optics and Lasers in Engineering 50, 248–255 (2012)

12. Nien, H.H., Huang, W.T., Hung, C.M., et al.: Hybrid image encryption using multi-chaos-system. In: 7th International Conference on Information. Communications and Signal Processing, ICICS 2009, December 8-10, pp. 1–5 (2009)
13. Prasad, M., Sudha, K.L.: Chaos Image Encryption using Pixel shuffling. In: Computer Science & Information Technology (CS & IT) CCSEA 2011, CS & IT 02, pp. 169–179 (2011)
14. Zhou, X., Ma, J., Du, W., Zhao, Y.: Ergodic Matrix and Hybrid-key Based Image Cryptosystem. International Journal of Image, Graphics and Signal Processing 4, 1–9 (2011)
15. Chen, L.: A Novel Image Encryption Scheme Based on Hyperchaotic Sequences. Journal of Computational Information Systems 8(10), 4159–4167 (2012)
16. Wang, X.Y., Yang, L., Liu, R., Kadir, A.: A chaotic image encryption algorithm based on perceptron model. Nonlinear Dynamics 62, 615–621 (2010)
17. Sakthidasan, K., Sankaran, Santhosh Krishna, B.V.: A New Chaotic Algorithm for Image Encryption and Decryption of Digital Color Images. International Journal of Information and Education Technology 1(2), 137–141 (2011)
18. Fu, C., Chen, J.J., Zou, H., Meng, W.H.: A chaos-based digital image encryption scheme with an improved diffusion strategy. Optics Express 20(3), 2363–2378 (2012)
19. Seyedzadeh, S.M., Mirzakuchaki, S.: A fast color image encryption algorithm based on coupled two-dimensional piecewise chaotic map. Signal Processing 92(5), 1202–1215 (2012)
20. Yáñez-Márquez, C.: Associative Memories Based on Order Relations and Binary Operators PhD Thesis. Center for Computing Research, México (2002) (in Spanish)
21. Acevedo, M.E., Yáñez, C., López, I.: Alpha-Beta Bidirectional Associative Memories: Theory and Applications. Neural Processing Letters 26, 1–40 (2007)

A Hypothetical Scenario-Based Analysis on Software Reliability Evaluation Approaches in the Web Environment

Jinhee Park, Hyeon-Jeong Kim, and Jongmoon Baik

Abstract. With the spread of the Internet and the development of Web technology, web-based software such as web applications and web services has been in the spotlight and widely used. Accordingly, ensuring web-based software reliability is becoming important, and the efforts to develop highly reliable software in the web environment are required. Compared with traditional software, research on the reliability of web-based software is not enough, and the dynamic execution environment of the web makes the reliability evaluation of web-based software much more complicated. In this paper, we deal with reliability evaluation issues in the web environment and compare with each other in terms of failure data collection methods, reliability evaluation techniques, and validation schemes. We also evaluate them based on hypothetical execution scenarios, analyze the strengths or weaknesses of each technique, and identify the remaining open problems.

Keywords: Reliability, web-based software, web application, web service, SOA system.

1 Introduction

With the changes of software development paradigms, web-based software has gained in popularity and is now providing various services to potential worldwide users [21]. In particular, the IT infrastructure of companies has been commonly implemented as the web-oriented system, and recently web services and Software as a Service (SaaS) delivering software via the web have been widely adopted for a business project. The web environment is easy to access and use, and it

Jinhee Park · Hyeon-Jeong Kim · Jongmoon Baik
Department of Computer Science, Korea Advanced Institute of Science and Technology, Deajeon, Republic of Korea
e-mail: {jh_park, jbaik}@kaist.ac.kr, hjkim@se.kaist.ac.kr

R. Lee (Ed.): *SNPD*, SCI 492, pp. 129–141.
DOI: 10.1007/978-3-319-00738-0_10

provides immediate feedback on software modifications. Likewise, the web environment is attractive to companies which attempt to reduce costs and has various potentials.

In spite of such benefits of the web-based software, it still suffers from various critical failures. The malfunctions of such software can endanger business opportunities and even ruin the reputation of a company. For example, eBay, one of largest online auctions, experienced a shutdown of the system, due to the database problem. It resulted in around $5 million loss and the damaged a relationship between the company and its customers [24]. Thus, it has become important to develop the web-based software with high quality. As a key attribute of software quality, software reliability, which means the probability that software operates without a failure for a specified time in a specified environment [11], is often concerned. It is a precondition for satisfying the client needs and realizing a potential profit.

Traditionally, software reliability growth models (SRGMs) for a monolithic application have been used to verify that software has achieved the reliability objective before release [8]. They provide an analytical framework for describing a reliability growth phenomenon as the residual faults of software are removed during operational testing [7]. As another major class of software reliability models, architecture based reliability models for component based software systems have been employed to predict the software system reliability. These models reflect the internal structure of software and are applicable after software architecture is created [19]. Although many studies have been performed to evaluate the reliability of traditional software, we cannot simply apply those models designed for the traditional software into the web-based software due to distinctions between two types of software. More specifically, compared with traditional software, it is difficult to obtain failure data for web-based software because of uncertainty in the external parties, and they are remarkably influenced by an unpredictable network condition. These characteristics make it much more complicated to estimate the reliability of web-based software. Although some techniques exist today to measure and assure the web-based software reliability, there still exist many new challenges and it requires new techniques with consideration of the characteristics of the web environment.

In this paper, we survey existing approaches to estimating the software reliability in the web environment, with particular emphasis on issues that are related to the web applications and web services. We compare existing approaches and identify strengths and weaknesses of each approach using scenario based evaluation. We also suggest the possible reliability challenges for end user programmers. Accordingly, this study will provide an insight into guiding the future research direction for web-based software reliability.

The paper is structured as follows. The motivating problems of this paper are discussed in Section 2. In Section 3, we review and discuss different reliability approaches in the web environment. Section 4 evaluates each technique using a scenario based evaluation and Section 5 lists remaining open problems. Finally, Section 6 concludes this paper.

2 Motivating Problems

Web-based software has unique characteristics such as complex network environment, huge user population, cross-platform support, and data security requirements [5]. According to a report using information from major companies such as Google and Microsoft [17], while reliability approaches for the traditional software concentrate on logical or computational errors of software systems, main causes of failures in the web environment are network errors, configuration errors, system overload, and resource exhaustion [20]. It means that, unlike the common belief that a software failure rate is constant without the modification of source code, the failure rate of web-based software may change according to network conditions or workload information. Accordingly, it is necessary for the reliability techniques for web-based software to reflect dynamic execution environment and web workload characteristics. Moreover, in the case of service-oriented architecture (SOA) systems, it is expensive to test remote services, which makes it difficult to obtain failure data. In such environments, the traditional solutions that software designers have adopted are no longer feasible [12]. Even though some authors claim that component based reliability techniques can be applied to SOA systems, there is no solid work [4], and the component based reliability approaches do not consider failure types in the web environment.

The heavy competition by the open network environment and the wider user population demands an immediate response to handle above difficulties in the reliability evaluation, and requires new reliability techniques to reflect the characteristics of the web environment. Nevertheless, relatively few reliability studies have been conducted for the software in the web environment. Therefore, we compare and analyze existing techniques in the categories of web application reliability and web service reliability.

3 Reliability Evaluation Techniques in the Web Environment

We survey existing techniques that are recent and prominent studies on the reliability evaluation of the web-based software, classify them according to target systems and applicable phases, and analyze them based on criteria which reflect characteristics of web-based software.

3.1 Web Application Reliability

3.1.1 Background

For web applications, the reliability is defined as the probability that the web operation is completed without a failure [22]. Failure sources of web applications are source contents, host, networks, browser, etc. In this circumstance, workload information is treated as an important factor for reliability evaluation.

3.1.2 Reliability Evaluation Techniques for Web Applications

Workload Measures from Server Logs. J. Tian et al. evaluate operational reliability for web applications [22]. They suggest four types of workload measures, such as bytes, hits, users, and sessions, to approximate an amount of time to use the applications. The failure data and system usage information are extracted from error logs in the servers, and these data are used to evaluate the web application reliability. Nevertheless, this approach is not applicable to some combination of the web workload and error codes [6]. Also, it just focuses on the web source content failures except the network failures, browser failures, or user errors.

SRGMs. In [26], authors use use an Yamada exponential testing effort function based model to evaluate web application reliability. Measures on an user session workload are treated as the usage time of web applications, and the failure data are obtained from log files of the server. [5] combines different SRGMs to evaluate reliability for web applications. In this approach, genetic algorithm (GA) technique is employed to give weight on each SRGM, and three weighted combinational models are used for reliability evaluation.

Markov Analysis. In [21], the reliability behavior of web applications is analyzed by Markov process. In other words, web pages that are functionally independent are regarded as the components of web applications, and their controls are shifted based on Markov process. Various combinations of web page reliabilities using a simulator are examined to evaluate the reliability of web applications. However, a lot of assumptions to build the Markov model weaken the practicality of the technique.

3.2 *Web Service Reliability*

3.2.1 Background

SOA systems are built upon bundles of web services, and consist of steps such as service publishing, discovery, composition, and execution. The basic working principle of the SOA systems is explained in [27], and existing reliability evaluation approaches for the SOA systems are simply described in [16].

The reliability evaluation techniques for SOA systems are mainly classified into atomic level analysis and system level compositional analysis. In Fig. 1, rectangles indicate input and output. The inputs show the ingredients needed to perform reliability evaluation for both level services. The ovals depict the steps of reliability evaluation for SOA systems, and the arrows represent the transitions between the steps and input and output information of the reliability evaluation.

The reliability value for an atomic web service (i.e., a basic web service unit that cannot be split further), is derived from failure information according to the number of service invocations. A composite service is a new service that is composed of multiple web services with structural relationships [10]. Most reliability evaluation approaches for a composite service use information such as a compositional

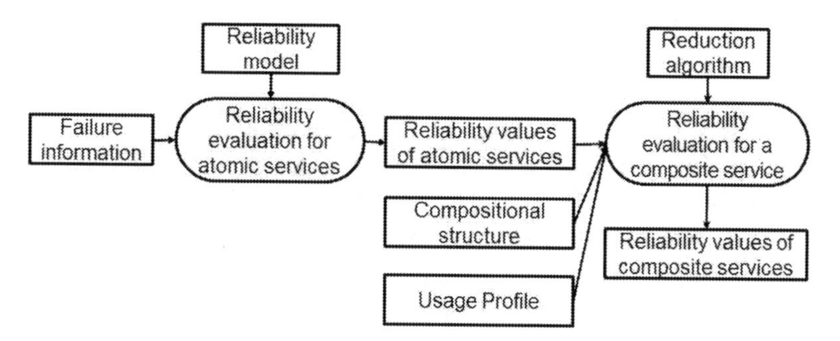

Fig. 1 The procedure for SOA system reliability evaluation

structure, usage profiles, and the reliability values of atomic services. The motivations of evaluating composite service reliability include the following:

- Figuring out how the reliability of atomic services and their structural information affect composite service reliability
- Helping to bind an appropriate service among multiple candidate services
- Selecting a compositional structure that is most appropriate for the SOA system

3.2.2 Reliability Evaluation Techniques for Web Services

Group Testing and Majority Voting. W. T. Tsai et al. present a testing based approach, which is called as service-oriented software reliability model (SORM) [23]. The reliability evaluation is performed by two steps. The first step is to perform group testing to obtain the service failure data. The services which generate the false output are detected by a majority voting technique, and testing results are recorded in error logs. The second step is to evaluate the atomic service reliability using error information based on outcomes of the voting.

Collaborative Framework. Zibin Zheng et al. propose a collaborative reliability prediction approach for atomic web services at the architecture design phase [28]. They predict the atomic service reliability using the information of other users who have similar experiences, and the aggregated failure probability of a service flow is derived from taking account of the compositional structures such as sequence, branch, loop, and parallel. This mechanism requires collecting failure data from different service users. With more accumulation of failure data, it is highly possible to obtain more accurate reliability prediction results for the web services.

Recalibration Technique. Andrew G. Liu et al. proposes a progressive reliability forecasting method [9]. They claim that reliability evaluation should be recalibrated because the web service reliability changes with time. This technique focuses on dynamic service execution circumstances such as shutdown of servers, network congestion, invalidity of data, and too many concurrent users. Since service failure patterns change as time goes by, this paper addresses when to (and how to) recalibrate the reliability to reflect the variation of the service reliability.

Multi-stage Reliability Model. In [25], authors state that the existing software reliability models cannot describe the failure occurred during the publishing, discovery, binding, and execution steps. Thus, a multistage model is proposed, which considers the possible web service faults in every step. The proper reliability model for each step is suggested and integrated for the service reliability evaluation.

Extended UDDI Model. B. Li et al. extend universal description, discovery, integration (UDDI) to monitor reliability-related information (e.g., the invocation, execution, and transition data) [10]. These data are consistently monitored and captured through the servers located in the UDDI register center. By using both the invocation and failure data, the reliability of atomic services is calculated by a Nelson model [14]. In this approach, log information from the extended UDDI is an important ingredient for the reliability analysis.

3.3 Comparison Criteria

One of major aims of this paper is to characterize some common practices in terms of web-based software reliability. Therefore, it is necessary to present comparison results among reliability approaches in the web. To do this, we review key papers, and establish five criteria. Table 1 summarizes the criteria to analyze reliability approaches in the web environment and their descriptions. Based on these criteria, we classify reliability approaches and perform a qualitative analysis of each approach.

Table 1 Criteria to analyze reliability approaches in the web environment

No	Criteria	Description
1	Target system	System domain on which proposed technique is applicable (e.g., SOA systems, web applications, etc.)
2	Phases	Applicable phases of corresponding technique
3	Failure data collection method	applied methods to obtain failure data
4	Technique	Used techniques or models for reliability estimation
5	Validation scheme	A criterion to check whether the proposed technique is validated or not

3.4 Overall Comparison Results

Table 2 shows the comparison results of existing reliability evaluation techniques for web-based software. Each technique is classified based on the target system and phase criteria. For web applications, failure information alone cannot characterize the software reliability because web workload information, which actively and dynamically changes, affects their reliability. Both the failure data and the web workload information are usually collected from web servers. Since existing SRGMs cannot describe the relationship between the workload and time, it is inappropriate to directly apply traditional SRGMs to web applications.

For SOA systems, most approaches express the service composition in a different way such as a Petri net model or a structure chart model, and they commonly use a stochastic workflow reduction (SWR) algorithm with some reduction rules to get the reliability value of a composite service [1]. This algorithm repeatedly applies six reduction rules, such as sequential, conditional, parallel, loop, fault tolerant, and network, to a workflow until only simple structure remains. Consequently, a major issue of reliability approaches in SOA domain is how to gather failure data to evaluate the reliability for an atomic service (i.e., determining service failure rate). We categorize existing reliability approaches according to the means of obtaining failure data which can be server logs, similar service users, or service testing in table 2. Through the comparison results, we can see that most approaches employ input domain reliability model (IDRM) such as the Nelson model [14] for calculating atomic service reliability. The IDRM is a model representing the probability that one of input data leads to a software failure as follows:

$$R = 1 - \frac{f}{n} = 1 - r \tag{1}$$

where f is the number of failures, n is the total number of service invocations (workload units), and r is the failure rate. This model has a hypothesis that faults are not immediately fixed after detecting faults. It provides a snapshot of current reliability and is often used at the end of a testing phase. The existing SRGMs are not used for atomic service reliability evaluation because the service focuses on the changes of service reliability during the dynamic execution rather than observing reliability growth during a testing phase.

In the view of a validation scheme, we classify a degree of validation into three levels: fully validated, validated through few experiments, and no validation. Studies with full validation have experiments to apply their approaches for a real world case; those with weak validation have results from simulation, or an illustrative examples; that with no validation does not have any experiment or case study at all (it just gives qualitative analysis of reliability). As shown in table 2, even though some approaches describe the techniques for web-based software reliability analysis, case studies have been rarely reported for these techniques. Most of them are performed under laboratory conditions [28, 10] or are simple examples [23, 9, 3, 20]. In addition, they usually have a small size, and consider only simple compositional structure cases for the SOA systems. It may avoid many practical problems in real systems. Industrial case studies for a large-scale web-based system with complex structures are urgent to validate the each technique.

Table 2 Comparison of reliability estimation techniques for web-based software

Target system	Phases	Failure data collection method	Citation	Reliability evaluation techniques		Validation scheme [a]
				Atomic service	Composite service	
Web application	Execution	Log information based (web server logs)	[22]	Nelson model		■ Real world experiment
	Testing, Execution	Simulation based	[21]	Markov Analysis		▣ Simulation
		Log information based (IIS server logs)	[5, 26]	NHPP model, genetic algorithm		▣ Case study
SOA system	Composition, Execution	Log information based (Extended UDDI model)	[10]	Nelson model	SWR algorithm (structure chart)	▣ Case study
		Collaborative approach (Similar user's information)	[28]	Exponential reliability function	SWR algorithm (structure chart)	■ Planet Lab
		Assumption of component reliability	[29, 3]	-	SWR algorithm (Stochastic petri net, Probabilistic flow graph)	▣ Example
		Testing based (Group testing)	[23]	Nelson model	SWR algorithm (Scenario model)	▣ Example
		Testing based (Recalibration technique)	[9]	Nelson model	-	▣ Case study
	Every SOA Steps	-	[25]	A staged reliability model	simple structure(AND, OR)	□ No validation

[a] ■ : fully validated, ▣ : validated through few experiments, □ : no validation

4 Evaluation

In this section, we evaluate the reliability techniques for web-based software using a scenario-based evaluation approach. To increase the accuracy of reliability evaluation, it is required for the techniques to cover changing situations in the web environment as much as possible. This evaluation approach describes how well each technique reflects dynamic execution scenarios. In this paper, we evaluate reliability techniques only for SOA systems because the SOA systems have the characteristics of both a web application and a distributed system. We generate the two types of hypothetical scenarios which influence web service reliability.

Scenario 1 a service provider tests the web service exhaustively and guarantee the defect free software based on Service Level Agreement (SLA). However, some clients experience failures due to following circumstances: (1) clients in an underdeveloped country have a trouble to get a seamless service due to the poor network facilities; (2) between 6PM and 8PM, a lot of service requests occur; (3) the server is sometimes down during the specific time in a day.

Scenario 2 a company wants to generate a new service. To do this, firstly, a web service composer searches the services to use through UDDI, and then builds the new service by composition of the selected services for a task. From this new service, (1) the execution probability of the each branch structure or the maximum number of repetition in a loop structure continuously changes. (2) service providers often update their service. Although the updated service is tested extensively, it affects the compatibility with other existing services in the composition.

The scenario 1 describes various situations to change the reliability of an atomic service. With this scenario, we can check if each technique can reflect the dynamic situations, such as unpredictable network condition (scenario 1-1), resource exhaustion (scenario 1-2), server availability (scenario 1-3). On the other hand, the scenario 2 describes the situations for a composite service such as usage profile changes (scenario 2-1) and service updates (scenario 2-2). With the scenario 2-2, we check dependency problems between services in the reliability evaluation. Table 3 shows the results of scenario based evaluation. Depending on how well each technique covers the scenarios, grades such as good, mediocre, or poor are manually appraised. The techniques are good or poor in some scenarios, and no perfect techniques reflecting all circumstances exist. It means that there is no technique that covers all

Table 3 Evaluation of the surveyed reliability estimation techniques

Methods[Citation]	Scenario 1[b]			Scenario 2[b]		Strength(+) and Weakness(-)
	1-1	1-2	1-3	2-1	2-2	
Extended UDDI model (Log information) [10]	□	▣	▣	▣	□	+ predict the reliability of a service before it is executed - possible to occur additional failures by UDDI overload
Collaborative technique [28]	■	□	▣	▣	□	+ robust to an unpredictable network problem - assume the availability of failure data from all the users - no complete reliability aggregation part (assumptions of branch selection probability and independent task failure in service flows)
Architecture based model[29, 3]	□	□	□	▣	□	- only applicable to web service composition - assume the availability of atomic service reliabilities
Group testing & Majority voting [23]	□	□	▣	□	□	- require the testing of functionally equivalent atomic services
Recalibration technique & Nelson model [9]	▣	■	■	□	□	+ robust to dynamic reliability changes - require much failure data before release
A staged reliability model [25]	▣	□	□	□	□	+ consider failures occurred in every SOA step - no suggestions of a failure data collection method

[b] ■ : Good, ▣ : Mediocre, □ : Poor

factors, which in turn affect accuracy on the reliability of entire services. Through this evaluation, the strengths and weaknesses of each technique are also analyzed. The recalibration and collaborative techniques are relatively proper to describe the dynamic web environment and to measure the service reliability during system operation, especially for an atomic service. We also found that no techniques afford scenario 2-2. We summarize findings from this scenario-based evaluation and the comparative analysis in the following section.

5 Remaining Open Problems

We list remaining open problems and give possible solutions from them.

Dynamic Execution Environment. Web services evolve quickly because service providers try to optimize the performance of services or add new functions to meet new user requirements [15]. Moreover, dynamic service binding, which helps to improve the system reliability, is often adopted for the fault tolerant SOA systems. Although such dynamic situations need to be captured and exploited for the reliability analysis, no works consider these situations. Through the evaluation section, we found that no approach covers all factors which affect the reliability of web-based software. Due to the dynamic web environment, it is difficult to build the reliability estimation reflecting all situations. Nevertheless, a general technique is required to accommodate as many situations as possible. Hybrid reliability approaches which combine various techniques might promise to handle most characteristics for the web-based software.

Assumptions of a Usage Profile. To compute the composite service reliability, structure information is used. In a branch or loop structure, the execution probability of the i-th branch, the value of maximum looping time, or the probability of executing the loop for i times is assumed to be provided by the system designer. However, it is difficult to identify these values before a service execution. Furthermore, if changes of the usage profile occur, the reliability value should be recalibrated. One solution to increase the accuracy of the values is to simulate the service flow multiple times and record the approximate values.

Dependency Problems between Services. Reduction rules of a composite web service by the SWR algorithm assume that web services are independent of each other. Most approaches fail to explain the dependency problems like scenario 2-2 in Section 4. These correlation problems need to be addressed in the near future. Some papers in the architecture-based reliability approach handle error propagation problems [18, 13, 2]. These papers can help to solve the dependency problems of the composite web service.

Determining Reliability of a New Service. In the case of a new service, it is difficult to obtain an accurate failure rate, because available information such as web log data or invocation information is sparse. Therefore, the reliability estimation of new services depends on the reliability value offered by a service provider. The service provider analyzes the reliability of the new service through a comprehensive test before a service publication. However, since an agile methodology based on open source software components, which has never considered in reliability studies, is often used to develop web-based software, reliability analysis for a new service is not an easy task, and reliability values from the viewpoint of service clients are ignored in this case.

Reliability Issues for End-User Programs. With the emergence of Web 2.0, web-based end-user programs such as Mash-up and Web-macro are developed by consumer-centric, which makes the best use of the consumer creativity for new software. In particular, Mash-up makes new services by reusing existing services, which is similar to a SOA system, and provides an easy way for service composition in the web environment. Although they also experience various failures such as network failures, server failures, and data source failures, no reliability research on this area exists. There is a plenty of room that the composite reliability estimation techniques for SOA systems can be applied to the Mash-up applications which are also the combination of external services.

6 Conclusion

In this paper, we analyze the representative papers in reliability research on web-based software, and evaluate them based on the hypothetical change scenarios of web service reliability. Through the evaluation, we identify the limitations of each approach and suggest the remaining problems of web-based reliability research. Although web-based development is in the spotlight, many challenges related to the reliability, such as insufficient validation, assumptions of a usage profile, and dependency issues, still remain. To the best of our knowledge, there has been no study on characterizing and evaluating reliability approaches in the web environment. Therefore, we expect that this paper will help to look through what is going on in this research field and to understand what is left for further research.

Acknowledgements. This work was supported by the National Research Foundation of Korea Grant funded by the Korean Government (MEST) (NRF-2010-C1AAA001-2010-001014375).

References

1. Cardoso, J., Miller, J., Sheth, A., Arnold, J.: Modeling Quality of Service for Workflows and Web Service Processes. Technical Report #02-002, LSDIS Lab, Computer Science, University of Georgia (2002)
2. Cortellessa, V., Grassi, V.: A modeling approach to analyze the impact of error propagation on reliability of component-based systems. In: Proceedings of the 10th International Conference on Component-based Software Engineering, pp. 140–156 (2007)

3. Grassi, V.: Architecture-based reliability prediction for service-oriented computing. In: de Lemos, R., Gacek, C., Romanovsky, A. (eds.) Architecting Dependable Systems III. LNCS, vol. 3549, pp. 279–299. Springer, Heidelberg (2005)
4. Grassi, V., Patella, S.: Reliability prediction for service-oriented computing environments. IEEE Internet Computing 10(3), 43–49 (2006)
5. Hsu, C.J., Huang, C.Y.: Reliability analysis using weighted combinational models for web-based software. In: Proceedings of the 18th International World Wide Web Conference, pp. 1131–1132 (2009)
6. Huynh, T., Miller, J.: Further investigation into evaluating website reliability. In: 4th International Symposium on Empirical Software Engineering, Noosa heads Australia, pp. 162–171 (2005)
7. Karanta, I.: Methods and problems of software reliability estimation, VTT Working Paper 63, Espoo (2006)
8. Kotaiah, B., Khan, R.A.: A survey on software reliability assessment by using different machine learning. techniques. International Journal of Scientific & Engineering Research 3(6) (June 2012)
9. Liu, A.G., Musial, E., Chen, M.H.: Progressive Reliability Forecasting of Service-Oriented Software. In: IEEE International Conference on Web Services, pp. 532–539 (2011)
10. Li, B., Su, Z., Zhou, Y., Gong, X.: A user-oriented web service reliability model. IEEE International Conference on Systems, Man and Cybernetics, 3612–3617 (2008)
11. Lyu, M.R. (ed.): Handbook of software reliability engineering. IEEE Computer Society Press (1996)
12. Lyu, M.R.: Software reliability engineering: a roadmap. In: Proceedings of FOSE, pp. 153–170 (2007)
13. Nassar, D.M., Shereshevsky, M., Gradetsky, N., Gunnalan, R., Ammar, H.H., Yu, B., Mili, A.: Error propagation in software architectures. In: Proceedings of 10th International Symposium on Software Metrics, pp. 384–393 (2004)
14. Nelson, E.: Estimating software reliability from test data. Microelectronics and Reliability 17(1), 67–73 (1978)
15. Nguyen, C.D., Marchetto, A., Tonella, P.: Test case prioritization for audit testing of evolving web services using information retrieval techniques. In: IEEE International Conference on Web Services, pp. 636–643 (2011)
16. Park, J.H., Baik, J.M.: Study on reliability approaches for SOA systems. In: Proceedings of 2013 Korea Conference on Software Engineering (January 2013)
17. Pertet, S., Narsimhan, P.: Causes of failures in web applications. CMU-PDL-05-109, Carnegie Mellon University (2005), http://repository.cmu.edu/pdl/48
18. Popic, P., Desovski, D., Abdelmoez, W., Cukic, B.: Error propagation in the reliability analysis of component based systems. In: Proceedings of 16th International Symposium on Software Reliability Engineering (ISSRE), pp. 10–62 (2005)
19. Goseva-Popstojanova, K., Mathur, A.P., Trivedi, K.S.: Architecture-based approach to reliability assessment of software systems. Performance Evaluation (45), 179–204 (2001)
20. Rahmani, M., Azadmanesh, A., Siy, H.: Architecture-based reliability analysis of web services in multilayer environment. In: Proceedings of the 3rd International Workshop on Principles of Engineering Service-Oriented Systems, pp. 57–60 (2011)
21. Suri, P.K., Bhushan, B.: Reliability evaluation of web based software. International Journal of Computer Science and Network Security 7(9) (September 2007)
22. Tian, J., Rudraraju, S., Li, Z.: Evaluating web software reliability based on workload and failure data extracted from server logs. IEEE Transactions on Software Engineering 30(11) (November 2004)

23. Tsai, W.T., Zhang, D., Chen, Y., Huang, H., Paul, R., Liao, N.: A software reliability model for web services. In: The 8th IASTED International Conference on Software Engineering and Applications, pp. 144–149 (2004)
24. Wang, W.L., Tang, M.H.: User-oriented reliability modeling for a web system. In: Proceedings of 14th International Symposium on Software Reliability Engineering (ISSRE), Denver, Colorado, pp. 293–304 (November 2003)
25. Xie, C., Li, B., Wang, X.: A staged model for web service reliability. In: The 35th Computer Software and Applications Conference, pp. 564–565 (2011)
26. Yang, J., Deng, Z., Wang, R., Hu, W.: Web software reliability analysis with Yamada exponential testing-effort. In: 9th international Conference on Reliability, Maintainability and Safety, pp. 760–765 (2011)
27. Zhao, S., Lu, X., Zhou., X., Zhang, T., Xue, J.: A reliability model for web services From the consumers perspective. In: International Conference on Computer Science and Service System, pp. 91–94 (2011)
28. Zheng, Z., Lyu, M.R.: Collaborative reliability prediction of service-oriented system. In: 32nd ACM/IEEE International Conference on Software Engineering, pp. 35–44 (2010)
29. Zhong, D., Qi, Z.: A Petri Net Based Approach for Reliability Prediction of Web Services. In: Meersman, R., Tari, Z., Herrero, P. (eds.) OTM 2006 Workshops. LNCS, vol. 4277, pp. 116–125. Springer, Heidelberg (2006)

On Distributed Energy Routing Protocols in the Smart Grid

Jie Lin, Wei Yu, David Griffith, Xinyu Yang, Guobin Xu, and Chao Lu

Abstract. The smart grid shall integrate the distributed energy resources and intelligently transmit energy to meet the requests from users. How to fully utilize the distributed energy resources and minimize the energy transmission overhead becomes critical in the smart grid. In this paper, we tend to address this issue and develop the distributed energy routing protocols for the smart grid. In particular, we first develop the *Global Optimal Energy Routing Protocol* (GOER), which efficiently distributes energy with minimum transmission overhead. Considering that the computation overhead of GOER limits its use in large-scale grids, we then develop the *Local Optimal Energy Routing Protocol* (LOER) for large-scale grids. The basic idea of LOER is to divide the grid into multiple regions and adopt a multiple layer optimal strategy to reduce the energy distribution overhead while preserving the low computation overhead. Through extensive theoretical analysis and simulation experiments, our data shows that our developed protocols can provide higher energy distribution efficiency in comparison with the other protocols.

Keywords: Smart grid, Distributed energy routing,Optimization, Energy distribution.

Jie Lin · Xinyu Yang
Xi'an Jiaotong University, P.R. China
e-mail: Dr.linjie@stu.xjtu.edu.cn, yxyphd@mail.xjtu.edu.cn

Wei Yu · Guobin Xu · Chao Lu
Towson University
e-mail: {wyu,clu}@towson.edu, tigerguobin@gmail.com

David Griffith
National Institute of Standards and Technology (NIST)
e-mail: david.griffith@nist.gov

R. Lee (Ed.): *SNPD*, SCI 492, pp. 143–159.
DOI: 10.1007/978-3-319-00738-0_11

1 Introduction

There has been an explosion of interest in the smart grid recently [1, 2]. As one type of cyber-physical systems (CPS) by integrating of physical and computational entities, with telecommunication technologies, the smart grid shall provide efficient, reliable, secure and resilient energy services [3]. In the smart grid, users can be provided with a reliable supply of power and efficient power production and resource utilization through monitoring and controlling the power transmission and distribution process. To achieve this vision, we shall not only develop innovative technologies in power transmission, storage and consumption, but also exploit techniques to effectively integrate distributed energy resources.

The development of power system and information technologies will increase the development and integration of distributed energy resources, which will augment and may eventually replace centralized power transmission as the technology evolves. Distributed energy generation, storage, and demand-side load management can change the way that we consume and produce energy [4]. Hence, the distributed energy management is an important tool for meeting growing energy demands through integrating renewable energy resources into the grid. A number of research efforts have been conducted in that direction [5, 6, 7, 8, 9, 10, 11, 13]. For example, Guan *et al.* [7] proposed minimizing the overall cost of electricity and natural gas for a building operation over a given time horizon. Jin *et al.* [9] proposed a system called Online Purchase Electricity Now (OPEN) to consolidate the customers' orders and determine an optimal generation and distribution scheme to satisfy customers' demands.

In this paper, we investigate the energy routing problem in the smart grid and propose a set of distributed energy routing protocols that determine power transmission paths based on energy requests from users and minimize the overall cost of power transmission. We formalize the problem of determining the power transmission paths as an optimization problem. Based on the formalized model, we first introduce the *global optimal energy routing protocol* (GOER). Considering that the computation overhead of GOER limits its usage in large-scale grids, we then develop the *local optimal energy routing protocol (LOER)* for large scale grids. The routing protocols collect the necessary information about grid, solve the optimization problem, and determine the amount of power to be transmitted through each link based on the supply and demand requests associated with users and the capacity limit of links, while minimizing overall resource consumption. We choose the energy transmission cost as the optimization objective; however, our solution is generic and can incorporate other resource optimization objectives.

We simulate the proposed protocols and compare them to a set of load-balancing protocols. We show that our approach achieves better performance and can provide reliable power with low transmission cost. We show that while the total power transmission cost of all protocols increases as the size of the grid increases, the total power transmission cost of our protocols is consistently lower than that of other protocols. The total power transmission cost of our protocols increases more slowly with respect to increasing energy-demand rate as well.

The remainder of the paper is organized as follows. In Section 2, we introduce the network model. In Section 3, we present the basic idea of our approach and the details of optimal energy routing protocols. We analyze the proposed energy routing protocols in Section 4. In Section 5, we show experimental results to validate the effectiveness of our approach. We offer conclusions in Section 6.

2 Network Model

The smart grid enables bi-directional flows of power transmission and data communication. With distributed energy resources (e.g., wind energy, solar energy, and others) and storage devices, each user can dynamically push power to or pull power from the grid to satisfy users' demand. In our paper, nodes are used to represent either energy demanders or suppliers. The grid elements are not homogeneous, i.e., each node in the grid generates and consumes some quantities of energy. Note that the supplied energy from nodes in a local area may correlate each other due to similar weather condition. When a node pulls energy from the grid, the node is denoted as an energy demand-node. When a node pushes residual energy into the grid, the node is denoted as an energy supply-node. Through energy transmission among energy supply-nodes and energy demand-nodes, the energy supply and demand can be balanced in the smart grid.

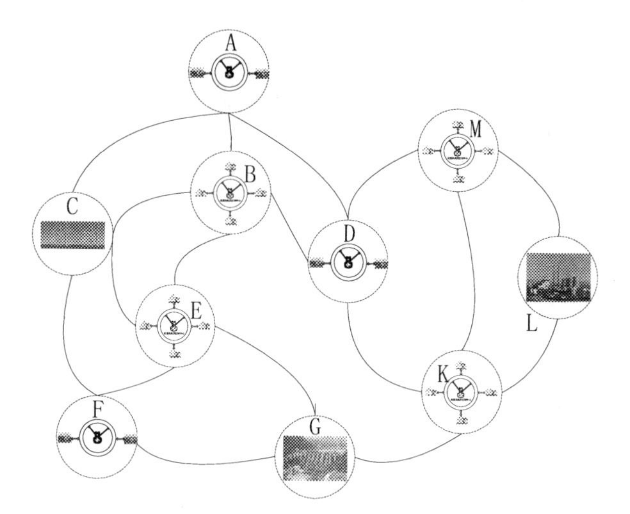

Fig. 1 The Structure of Smart Grid

As shown in Fig. 1, we consider two types of links. Particular, the communication links used to interactively exchange the energy demand and supply messages among nodes. The energy links are used to transmit power from supply-nodes to demand-nodes. As we can see, these two types of links integrate all nodes into the grid and enable effective use of the distributed energy resources and the balance of energy demand and supply in the grid.

Each user can generate power through distributed energy resources (e.g., solar energy and others) and store the generated power locally. All electric appliances associated with users can obtain power from the energy-storage equipments. Measuring components (e.g., smart meter and other devices) can be used to measure the energy consumption that is used by electric appliances, the energy generation and the energy storage.

We assume that the measuring components can determine whether a node belongs to a supply-node or demand-node. To balance demand and supply, each node communicates with other nodes, sharing measurements, energy availability, and requests for energy. We define an energy transmission cycle as a period of time in which the supply and demand states of nodes in the grid do not change significantly. Denote $E^i_{supl}(n)$ as the difference between the supplied and demanded energy at the ith node during the nth energy transmission cycle. If $E^i_{supl}(n) > 0$, the ith node can provide at most $E^i_{supl}(n)$ units of energy to the set of demand-nodes. If $E^i_{supl}(n) \leq 0$, the ith node requires $|E^i_{supl}(n)|$ units of energy from the set of supply-nodes during the nth energy transmission cycle.

We also assume that the energy transmissions between supply-nodes and demand-nodes are scheduled periodically over energy transmission links. During each energy transmission cycle, each energy-link can transmit energy in one direction only. The unit transmission cost on an energy-link that connects nodes i and j is

$$Cost_{ij} = a \cdot \ell_{ij} \cdot E, \tag{1}$$

where a is average cost per mile, ℓ_{ij} is the link's length, and E is the unit of energy transmitted over the link (e.g., 1 Megawatthours) [14], and $Cost_{ij}$ is the cost of transmitting one unit energy on the link and its unit is $\$M$ ($ per Megawatthours).

3 Our Approach

In this section, we first introduce the basic idea of our approach and present the detailed design of energy routing protocols.

3.1 Basic Idea

Recall that the energy generation and consumption process of the nodes in the smart grid may be unbalanced. To resist the unbalance, some nodes can be supply-nodes and provide extra energy to demand-nodes. Nevertheless, if multiple demand-nodes exist in the smart grid, they would affect each other to obtain energy on time, leading to high resource consumption. To address this issue, we propose the distributed energy routing protocols. In particular, we develop the *global optimal energy routing protocol (GOER)* and the *local optimal energy routing protocol (LOER)*, which can determine the optimal energy routes for transmitting power among nodes. Based on the optimal energy routes, supply-nodes can provide the reliable supply of power to all demand-nodes in the smart grid, and the minimum resource (e.g., energy transmission cost) will be used due to energy transmission.

As the smart grid consists of multiple supply-nodes, demand-nodes, and energy transmission links and different links between the nodes have different capacity, we consider the constraints of power grid and formalize the problem of selecting the optimal energy routes as an optimization problem. We develop the energy routing protocols to determine the quantity of power transmitted through links to balance energy supply and demand along with link capacity while minimizing overall transmission cost.

Our distributed energy routing protocols consist of two sub-protocols: *the global optimal energy routing protocol (GOER)* and *the local optimal energy routing protocol (LOER)*. In GOER, we consider the energy transmission in the smart grid as a global optimization problem. By solving the problem, GOER can determine the amount of power to be transmitted through the individual links and then derive the optimal energy routes to guide the power transmission among nodes, while the overall transmission cost is minimized. Hence, GOER is a perfect protocol to select optimal energy routes. However, the computation overhead of GOER increases rapidly when the size of grid increases. Hence, we propose LOER to divide the large grid into multiple regions. In LOER, a multiple-layer optimization is conducted. That is, the optimal energy routes are conducted first in individual regions and the optimal energy routes are then performed in across regions based on GOER. Although LOER will not derive the optimal energy routes as GOER does, LOER balances resource consumption and computation overhead. In LOER, we consider the following two ways to construct regions: In a static region, the grid is pre-divided into regions with different levels. In a dynamic region, the grid is divided based on the power-demands from nodes, and lower-level regions can compose a higher-level region dynamically.

3.2 Global Optimal Energy Routing Protocol (GOER)

Recall that in the smart grid, all nodes can be categorized into two groups: supply-nodes and demand-nodes and grid can be formalized as a graph as shown in Fig. 1. The problem is to determine the routes for transmitting energy from supply-nodes to demand-nodes, leading to the minimum cost of power transmission. We propose a global optimal energy routing protocol (GOER), which can not only provide the reliable supply of energy to all demand-nodes, but also consume the minimum resource while delivering energy. Once again, in this paper, we choose the energy transmission cost as the optimization objective. Nevertheless, our solution is generic and other types of resource optimization objectives can be applied as well.

Our routing protocol needs to consider the following three constraints: *Constraint 1:* The output energy from the node plus the demanded energy at that node should be equal to the input energy of that node. *Constraint 2:* The input energy plus the generated energy on the node should be no less than the output energy from the node. *Constraint 3:* The amount of energy transmitted on a link should not exceed the load-carrying capacity of link.

The energy input into a node or output from a node should be transmitted through the link directly connected to that node. The supply-node does not care about where the energy will be transmitted and the demand-node does not care where the received energy came from originally. We only need to determine the amount of energy that each link should transmit based on the above constraints. As we can see, we transform the problem of how to optimally transmit the energy among nodes into a global optimization problem. By solving this global optimization problem, we obtain the optimal energy routes to guide the energy transmission between supply-nodes and demand-nodes. The details of our algorithm are described below.

First, each demand-node broadcasts the request message with the amount of demanded energy to all nodes in the grid. After receiving the request message, these nodes send the link-state information of all the connected links back to these demand-nodes. The link-state information of a link is represented as: (***ID, NID, E_{rest}^{ID}, Length, Load, Cost, Level***). When the demand-node receives the link-state information of all links in the grid, it determines whether it requests the largest amount of energy in the grid. If it does, it becomes the master node in the grid. Otherwise, it is not the master node and does nothing. Note that the master node should only be the demand-node, which is elected from all demand-nodes dynamically in each cycle, and can determine the energy to be transmitted for each link.

We now have the master node process the link-state information of all energy-links in the grid. It treats the amount of energy that each link should transmit as an unknown parameter (e.g., either positive or negative number), and establish the optimization function of the power transmission. Note that, the amount of energy to be transmitted through the individual links should satisfy the constraints stated above. The energy transmission cost should be the minimum as well. According to the above analysis, the problem of selecting the best energy routes can be formalized as an optimization problem that determines the amount of energy transmitted through the set of energy links based on the supply and demand requests of all the nodes along with the set of energy link capacity constraints, so that the overall transmission cost is minimized. The formalization of GOER is listed below.

$$\textbf{\textit{Objective.}} \quad Min\left\{ Cost = \frac{1}{2} \cdot \sum_{L_{ij} \in L} (|E_{ij}(n)| \cdot Cost_{ij}) \right\}$$

$$\textbf{\textit{S.t.}}$$

$$\begin{cases} \forall v \in N_S, \ \sum\limits_{i \in N_v} E_{vi}(n) \leq E^{v}_{\text{supl}}(n) \\ \forall u \in N_D, \ \sum\limits_{j \in N_u} E_{uj}(n) = E^{u}_{\text{supl}}(n) \\ \forall L_{ij} \in L, \ E_{ij}(n) = -E_{ji}(n) \\ \forall L_{ij} \in L, \ |E_{ij}(n)| \leq Load_{ij} \end{cases}, \quad (2)$$

where i and j are the IDs of nodes, $Cost_{ij}$ is the transmission cost on energy link L_{ij} connecting nodes i and j, $E_{ij}(n)$ is the energy transmitted on energy link L_{ij}, L is the set of all energy links in the grid, N_S is the set of supply-nodes, N_D is the set of demand-nodes, N_v and N_u are the sets of neighbor-nodes of node v and node u, respectively, and $Load_{ij}$ is the capacity of link L_{ij}. Note that the above formalization was briefly mentioned in our previous work [12], which is mainly focus on the false data injection attacks in energy routing. Differently, this paper is focus on developing effective energy routing protocols.

By solving the optimization problem, the master node obtains the amount of energy that each link needs to transmit, and the direction of transmission. Based on the problem formalization listed here, in Section 4, we formally analyze that GOER can satisfy the demands from all requests with minimum resources consumption during energy distribution.

3.3 Local Optimal Energy Routing Protocol

When GOER is used for a large grid, it becomes difficult to obtain the optimal energy routes with limited computation resources because the computation overhead increases rapidly with the size of the grid. To address this issue, we propose the *local optimal energy routing protocol* (LOER) for large-scale grids. LOER will not be able to derive the optimal energy routes as GOER does, but it can balance the resource consumption and computation overhead.

The main idea of LOER is to first divide the large grid into multiple regions, then obtain the optimal energy routes in each region based on GOER, and finally derive the suboptimal energy routes based on the obtained multiple optimal energy routes. A new region can consist of multiple low-layer regions and the new region would be a high-layer one. A link belongs to a region only if both nodes (or regions) it connects to belong to that region.

Multiple layer optimization is conducted in LOER. The optimal energy routes are conducted first in the higher-layer region and the optimal energy routes are then conducted in the associated lower-layer regions of the higher-layer region based on GOER, respectively. Note that, when we use GOER in a higher-layer region, the composed lower-layer regions are considered as

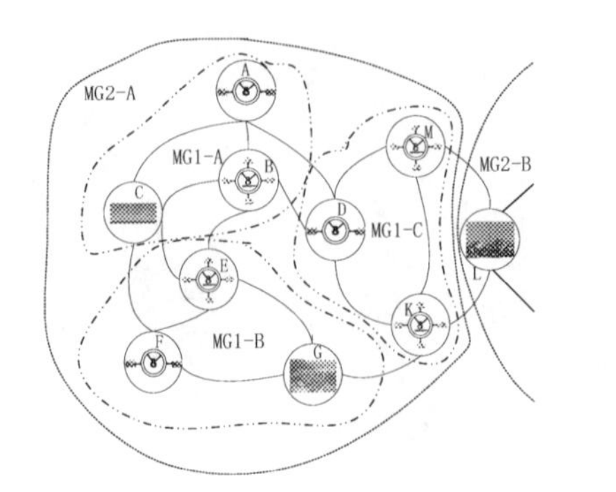

Fig. 2 Establishing Static Regions

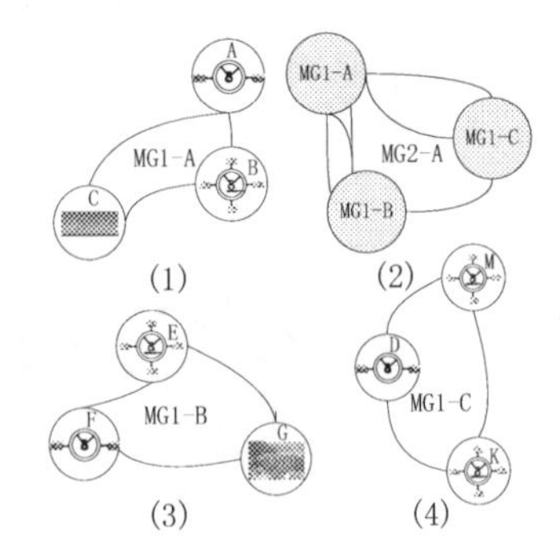

Fig. 3 Example of Static-LOER

nodes. Then, the higher-layer region can be considered as a grid as we defined in Section 2, and can use GOER to derive the optimal energy routes to guide the power transmission among low-layer regions. A region can be denoted as an energy-balance region if supply-nodes in this region can provide enough power for demand-nodes in this region.

According to the main idea of LOER mentioned above, the key issue is how to divide the global grid into multiple regions with multiple levels and select the energy-balance regions. To deal with it, we propose two mechanisms to divide the smart grid: static region and dynamic region based mechanisms. LOERs based on these two region establish mechanisms are denoted as static-LOER and dynamic-LOER, respectively. The details of these two protocols are described below.

3.3.1 Static-LOER

Static-LOER divides the global grid into multiple regions during the initialization phase. Note that the region assignment can be based on the geo-location of nodes. Each region is assigned a numerical level, and a $level_i$ region consists of at most N $level_{i-1}$ regions, where N is determined by the computation capacity of the node in the grid. Obviously, a $level_0$ region is a single node. As shown in Fig. 2, with the size of each region being 3, nodes A, B, C, nodes E, F, G, and nodes D, K, M compose the regions $MG1-A$, $MG1-B$ and $MG1-C$, respectively. These three $level_1$ regions compose the $level_2$ region $MG2-A$.

Because Static-LOER uses GOER in each region, each link in the grid needs to be assigned a link-level. In Static-LOER, we define the link-level of an energy-link to be one less than the level of the lowest-level region that contains the entire link. As shown in Fig. 2, the link-level of link L_{AB} is $level_0$ since it connects two $level_0$ nodes within a $level_1$ region, and the link-level of link L_{CF} is $level_1$ because it connects two $level_1$ regions.

The link-level of each link is set during the initialization phase. Based on these pre-established static regions, Static-LOER first selects the energy-balance regions, which are regions whose total demand is less than their total supply. Static-LOER derives the optimal energy routes among the low-level regions in the energy-balance region and derives the optimal energy routes within each low-level region. Finally, based on multiple derived optimal energy routes, Static-LOER selects the suboptimal energy routes to guide the energy transmission among nodes. With Static-LOER, all demand-nodes will be supported by supply-nodes with less computation cost and more consumed resources in comparison with GOER.

We now use Fig. 2 to illustrate how Static-LOER works. We assume that only nodes A and B are demand-nodes. Both nodes send the request message to other nodes via links at $level_0$, i.e., sending the message to other nodes at $level_1$ region which node A and node B belong to. Assume that region $MG1 - A$ is not an energy-balance region, i.e.,

$$|E^A_{\text{supl}}| + |E^B_{\text{supl}}| > E^C_{\text{supl}}. \tag{3}$$

Then, nodes A and B need to obtain energy from other $level_1$ regions, such as $MG1-B$ and $MG1-C$. These regions, along with $MG1-A$, compose the $level_2$ region $MG2-A$. If $MG2-A$ is an energy-balance region, Static-LOER first derives the optimal energy routes among $MG1-A$, $MG1-B$, and $MG1-C$, shown in Fig. 3-(2). Note that, at this moment, the power demanded by $MG1 - A$ is $|E^A_{\text{supl}}| + |E^B_{\text{supl}}| - E^C_{\text{supl}}$. Static-LOER then derives the optimal energy routes within $MG1 - A$, $MG1 - B$ and $MG1 - C$ based on GOER, shown in Fig. 3-(1), Fig. 3-(3), and Fig. 3-(4), respectively. By combining the derived optimal energy routes, Static-LOER obtains the suboptimal energy routes for node A and node B in $MG1 - A$. If, after optimization at $level_1$, $MG2 - A$ is not an energy-balance region, then $MG2 - A$ is treated as a node that the algorithm combines with other $level_2$ regions to compose a higher-level region. The process continues until an energy-balance region is produced; the highest level region is the global grid itself. Hence, with the assumption that the amount of total generated power is more than the total amount tha is consumed, Static-LOER can always select an energy-balance region for demand-nodes, and derives suboptimal energy routes to guide the reliable supply of power for demand-nodes in the grid.

3.3.2 Dynamic-LOER

Dynamic-LOER organizes the nodes into regions dynamically at the beginning of each energy transmission cycle. In Dynamic-LOER, when a demand-node needs to download energy from supply-nodes, it first sends a request message to all its neighboring nodes. Based on the response, the requesting node determines the number of demand-nodes and the number of supply-nodes among its neighbors. If a neighboring node requests more power than

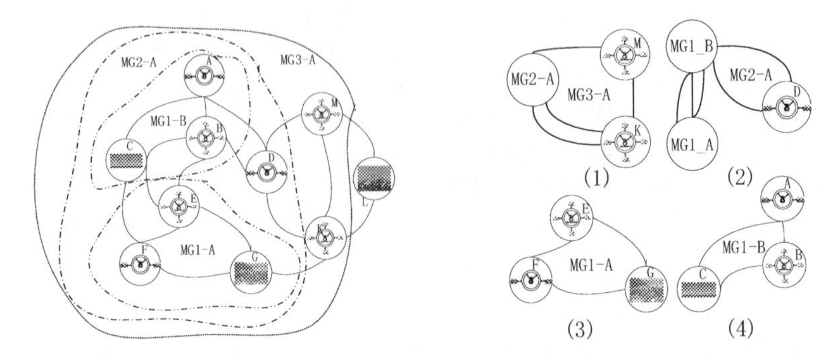

Fig. 4 Establishing Dynamic Regions **Fig. 5** Example of Dynamic-LOER

the requesting demand-node, the neighboring node will be set as the master node. Otherwise, the requesting demand-node will be set as the master node. Only master nodes can send region-establishing requests to perform dynamic region construction. The purpose of this is to reduce the collisions of region-establishing requests from neighboring nodes.

The details of constructing regions are described below. The master node and its neighboring nodes compose a new region. To control the computation overhead, similar to Static-LOER, the maximum size of a region in Dynamic-LOER is defined as N as well. If the number of neighboring nodes of the master node is larger than $N-1$, the master node randomly selects $N-1$ of notes to establish the new region. Otherwise, all of the master node's neighbors are included in the new region. In dynamic-LOER, a newly established region is set to region-level $level_{i+1}$, where $level_i$ is the highest level of the regions that compose the new region. As shown in Fig. 4, region $MG1-A$ and $MG1-B$ are both at $level_1$ and $MG2-A$ is at $level_2$. As defined in Static-LOER, each node in the grid is considered a $level_0$ region. After a region is established, it can be considered as a node in a network of peer regions. If a node receives multiple region-establishing requests, it joins the region with the lowest level.

Similar to Static-LOER, Dynamic-LOER uses the link-level to associate links with regions, and the link-level of an energy link is defined as the higher level of two nodes or regions that the link connects to. For example, the link-level of link L_{BD} in Fig. 4 is $level_2$ because D is in $MG2-A$ while node B belongs to $MG1-B$ and $level_1$.

The process of constructing regions stops when the newly established region is an energy-balance region. After determining the energy-balance region, Dynamic-LOER obtains the suboptimal energy routes in the same fashion as Static-LOER. For example, as shown in Fig. 4, we assume that node A and node F are demand-nodes. According to Dynamic-LOER, the final region will be $MG3-A$ if it is an energy-balance region. Then Dynamic-LOER derives the optimal energy routes among $MG2-A$, M and K in $MG3-A$ based on GOER, shown in Fig. 5-(1). Dynamic-LOER then derives the optimal energy routes among $MG1-A$, $MG1-B$ and D in $MG2-A$ based on GOER, shown in Fig. 5-(2). Finally, Dynamic-LOER derives the optimal energy routes in region $MG1-A$ and $MG1-B$ based on GOER, respectively, as shown in Fig. 5-(3) and Fig. 5-(4). Via combining these derived optimal energy routes, Dynamic-LOER obtains the suboptimal energy routes to guide the reliable supply of power for node A and node F, and close optimal power transmission cost and less computation overhead could be achieved.

4 Analysis

In this section, we analyze the effectiveness of our proposed algorithms, including the minimum transmission resource consumption, computation overhead, and storage overhead.

4.1 Minimum Resource Consumption

In the smart grid, where the amount of generated power is larger than that of consumed power, the selection of energy distribution routes can be conducted by different schemes to provide the reliable energy distribution to the demanders. Nevertheless, the routes selected by our GOER not only achieve this objective, but also lead to the minimum resource consumption during the power distribution. We have the following Theorem 1.

Theorem 1. *The energy routes selected by GOER can satisfy the demands from all requesters with minimum resource consumption during energy distribution.*

Proof. Let the energy route selected by GOER protocol be $\mathbf{R}$. If Theorem 1 is not true, the energy transmitted on $\mathbf{R}$ does not incur minimum transmission overhead. Then there is another route $\mathbf{R}'$, such that the resource consumption cost caused by $\mathbf{R}'$ is lower than the one caused by $\mathbf{R}$. Then, we have $\boldsymbol{Cost}(\mathbf{R}') > \boldsymbol{Cost}(\mathbf{R})$. In order to not conflict with the optimization function (e.g. Equation (2)) in our GOER, which selects an energy route with the minimum resource consumption cost, $\mathbf{R}'$ can only be an energy route, which does not satisfy the constraints of the optimization function. However, the constraints of the optimization function reflect the reliable energy supply to all demanders. Hence, $\mathbf{R}'$ cannot ensure that all requesters are provided with enough energy. This will conflict with the objective of $\mathbf{R}'$ to provide energy supply to all demanders. Hence, Theorem 1 is proved.

Note that, in Theorem 1, we validate that the GOER obtains the energy distribution routes with the minimum resource consumption. However, because of using GOER in a large grid will incur a high computation overhead on individual nodes, LOER is developed to balance the computation overhead and the resource consumption for large-scale grids. Hence, the energy distribution routes obtained through LOER may not be optimal, but they lead to low resource consumption with acceptable computation overhead. In Section 5, our experimental data shows that the resource consumption of energy distribution routes obtained through LOER is lower than that of other protocols.

4.2 Computation Overhead

Recall that the main idea of GOER and LOER is to transform the problem of determining the optimal power transmission paths as an optimization problem, in which multiple source nodes (i.e., a node can be treated as as a user), multiple destination nodes, multiple routes are considered, and links connected to nodes with capacity constraints. By solving the optimization problem, our protocols can obtain the desired routes for transmitting energy. Solving the optimization problem incurs the computation overhead as well. The optimization function is non-linear programming (NLP) with linear constraints and non-linear objective functions. Numerous approaches have been developed to solve this problem, and these approaches can be categorized into two types: traditional methods (e.g., SQP [16]) and deductive methods (e.g., simulated annealing [15], and genetic algorithm [17]). In addition, the non-linear programming can be transformed as the linear programming. In the following, we use the simulated annealing [15] as an example to analyze the computation overhead of GOER and LOER.

When the simulated annealing is used to solve the optimization problem, two rounds (the outside and inside) are included. The outside round controls the number of iteration of inside circle and can stop the algorithm, while the inside round selects a set of solutions for the optimization problem and studies the difference of the values of the objective function with these solutions. In our case, the simulated annealing to solve the optimization function will have the following steps: (i) Solving the m unknown parameters with m equations, which consist of n unknown parameters in total, where n and m are the number of the energy links and the energy-demanding nodes in the grid, respectively. Note that we only show the computation overhead caused by GOER and the overhead caused by LOER can be derived in the similar manner. (ii) Initializing the n unknown parameters. (iii) Repeating outside round M times, including L times of inside round in each time. (iv) Checking the constraints of optimization function P times and concluding two values of objective function in each inside circle. (v) l functions should be concluded while checking a constraint of optimization function, where l is the number of energy-provider nodes in the grid.

After the above five steps are completed, the optimization function will be solved and the energy routes will be derived. The computation overhead can be denoted as,

$$T(n, m, l) = (m - 1)n + n + ML(2 + Pl), \tag{4}$$

where $(m-1)n$, n, $ML(2+Pl)$ are the computation overhead incurred in steps (i)-(ii) and steps (iii)-(v). As we can see from Equation (4), the computation overhead increases with the increase of m, l, n, i.e., the size of grid. To increase the accuracy of the solutions, parameters M, L, and P, (defined above), are selected as large integer values. Hence, if the size of grid is large, the computation overhead will be high and the energy routes will not be computed on time. To deal with this, our proposed LOER reduces the computation overhead by dividing the large grid into multiple regions and optimization is conducted in individual regions and across regions. According to LOER, the computation overhead of LOER becomes

$$T = \sum T(n_i, m_i, l_i) = \sum (m_i n_i + ML(2 + Pl_i)), \quad (5)$$

where n_i, m_i, and l_i are the number of level i links, demand-nodes, and supply-nodes, respectively. Because the $\sum l_i < l$, and $\sum (m_i + n_i) < (m+n)$, the computation overhead of LOER is obviously less than that of GOER.

4.3 Storage Overhead

As we stated in the description of GOER and LOER in Section 3, a node in the grid should store the link state information of all links that it connects to, including

$$(\boldsymbol{ID}, \boldsymbol{NID}, \boldsymbol{E}^{\boldsymbol{ID}}_{\boldsymbol{rest}}, \boldsymbol{Length}, \boldsymbol{Load}, \boldsymbol{Cost}, \boldsymbol{Level}).$$

We assume that the average number of the links to which each node connects is n'. When a node is selected as a master node used to compute optimal energy routes, it should compute the energy routes for the grid. In this regard, the master node should collect the link state information of all transmission links. Note that, if the LOER is used, it only collects the link state information of the links within the region, where the master node tracks the supply and demand of energy in the region. Obviously, in an extreme case, the region should be the gird itself. Hence, no matter either using GOER or LOER, a master node should collect the link state information of all energy links in the grid. Hence, the storage overhead of the master node becomes

$$SO = n \cdot Size((\boldsymbol{ID}, \boldsymbol{NID}, \boldsymbol{E}^{\boldsymbol{ID}}_{\boldsymbol{rest}}, \boldsymbol{Length}, \boldsymbol{Load}, \boldsymbol{Cost}, \boldsymbol{Level})), \quad (6)$$

where n is the number of links in grid, and *Size* is a function for obtaining the size of parameters in brackets. We assume that the size of each parameter in the link state information is 64 bits, and the number of energy links in a grid is 1000. Then the maximum storage overhead of node can be $(1000 * 9 * 64)$ bits, which will only occupy a small chunk of memory. Hence, the storage overhead of our proposed protocols (i.e., GOER and LOER) is low for individual nodes in grid.

5 Performance Evaluation

In this section, we present the simulation results of our GOER and LOER protocols in comparison with other protocols with different objectives (e.g., load balance) in terms of the total power transmission cost.

5.1 Evaluation Methodology

To demonstrate the effectiveness of our developed protocols, we conduct simulations based on the topology of the US power grid [18]. To simplify the simulation study, we selected the major city of individual states as one node in the topology. The backbone of the interstate power transmission is based on the connection between these nodes. The fifty US states were selected as simulation objects, which were divided into five regions during the simulations. To simulate LOER, each region has two lower level sub-regions.

The data used for the simulation is based on "2009 US Energy Information Administration State Electricity Profiles" [19]. The averaged real-time data per second on each link is computed based on the averaged yearly data. The maximum power transmission of the link represents the capacity of the link in our experiments. The length of the link which represents the distance between two paired nodes is calculated using Google Maps.

To evaluate the effectiveness of our algorithms, we consider two load balance algorithms: (i) the local load balance algorithm, and (ii) the global load balance algorithm. These use different objective functions and tend to reduce equipment costs by minimizing the variation of power transmission on all links. Note that higher transmission capacity on the link results in a higher cost. This algorithm can balance the load on links in the premise of satisfying demand. The evaluation data below shows that our GOER and LOER can perform better than these load balance algorithms.

Because the purpose of the protocols is to provide reliable energy for the grid, we want to minimize the total power transmission cost. Because all the algorithms that we simulated can achieve this goal, our evaluation data focuses on the total power transmission cost versus various factors: the grid size, the amount of energy requested from demand-nodes, and the capacity of the links.

Following the cost definition in Section 2, we assume that the link cost parameters are $a = 2.6$ [14]. The data transmission cost between two nodes is given by Equation (1). Note that the power input or output of a node should be limited to the power that is requested or generated. Recall that our protocols are designed to select energy routes where the transmitted power must meet the link capacity constraint, demanded power, and supplied power, while the total power transmission cost, which is the sum of the cost on all links, is minimized. In all figures in this section, the unit of power transmission cost is $\$MM$ ($ per Million Megawatthours).

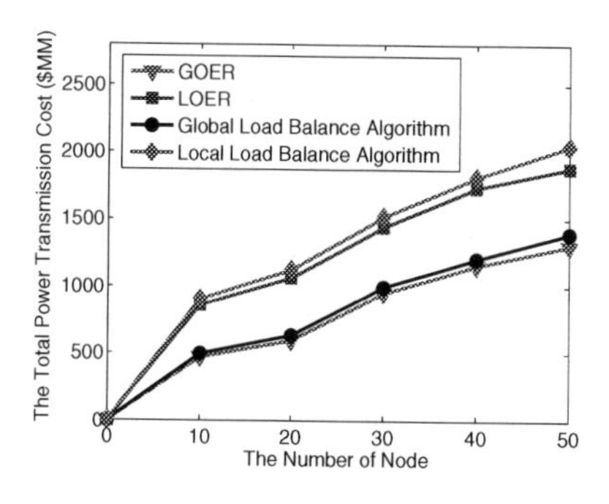

Fig. 6 Number of node vs. Power Transmission Cost

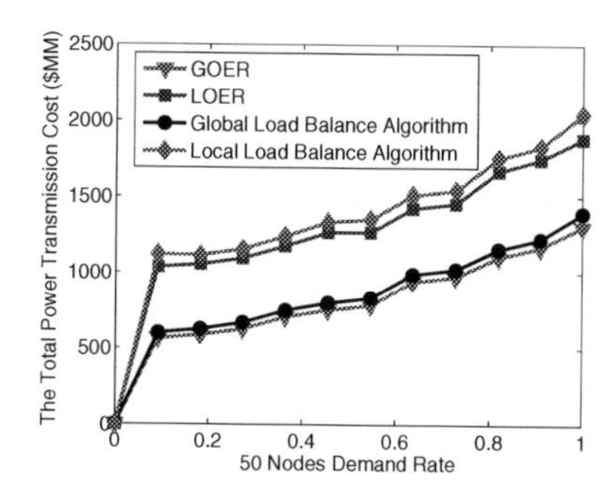

Fig. 7 Power Transmission Cost vs. Demand Rate

Note that, in our simulation, the non-linear programming is transformed to a linear programming and solved by built-in functions of Matlab 7.0. All simulations in this paper were carried out using Matlab 7.0. Note that, for LOER, we considered two grid-dividing mechanisms: static and dynamic. Because the dynamic mechanism to establish regions can be treated as an extension to the static mechanism, the performance of Static-LOER is similar to that of Dynamic-LOER.

5.2 Evaluation Results

Fig. 6 illustrates the relationship between the total power transmission cost and the number of nodes. In our simulation, we used only 10 nodes that correspond to the western states of the US. The nodes in the topology are arranged from west to east. As shown in this figure, the cost of GOER is less than the cost of the global load balance algorithm, which shows that our algorithm is optimized in terms of resource consumption caused by power transmission. With the increase in the number of nodes, the cost increases approximately in a linear fashion. The different curves for both the global algorithms and local algorithms demonstrate that in same grid, the cost of GOER is much smaller than LOER.

Fig. 7 depicts the impact of the demand rate on the total power transmission cost. Note that the demand rate is defined as the ratio of the number of demand energy nodes in the total number of nodes. The local schemes do not do as well as the global algorithms, because they optimize only subsets of the grid. Over the whole range of demand rates, the GOER and LOER algorithms performed better than their load-balance counterparts.

Fig. 8 depicts the total power transmission cost vs. the load capacity of links. We varied the link capacity from 3600 MW to 5000 MW. When the load is less than 3600 MW, there is not enough capacity to meet the energy

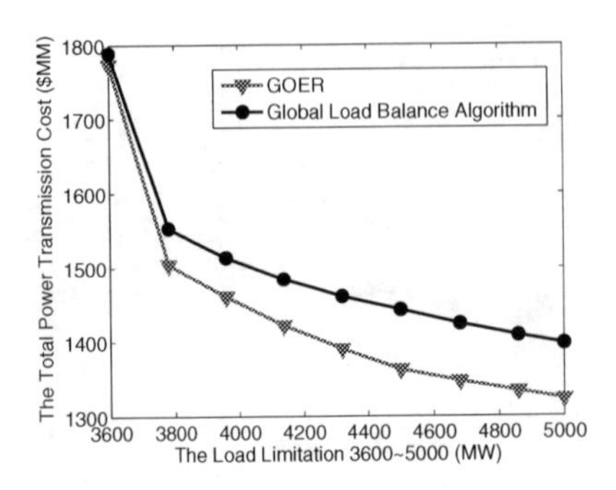

Fig. 8 Power Transmission Cost vs. Capacity Limitation of Link (3600 MW-5000 MW)

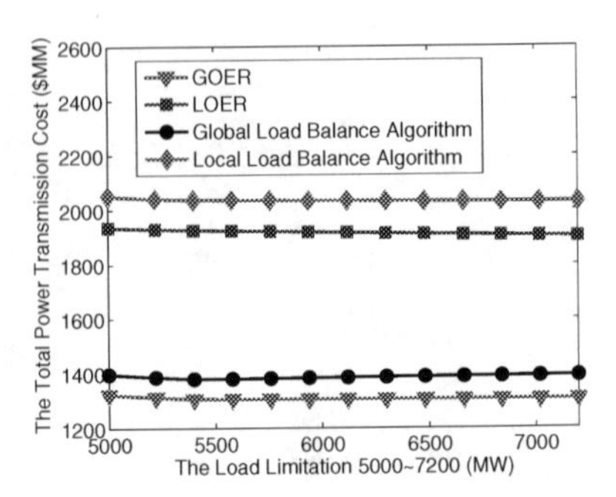

Fig. 9 Power Transmission Cost vs. Capacity Limitation of Link (5000 MW-7200 MW)

demand in our data set. When the capacity is more than 5000 MW, the result is almost the same as the case where no load capacity limitation exists. As we can see from this figure, the load capacity limit from 3600 MW to 5000 MW has a great influence on the total power transmission cost. Larger capacity links can benefit from the reduction of the power transmission cost by using our energy route optimization scheme.

Fig. 9 depicts the performance of these four algorithms as the link capacity limitation ranges from 5000 MW to 7200 MW. From Fig. 8, we can see that when the capacity is less than 5000 MW, the LOER cannot make nodes obtain enough power. With the increase of the link capacity, i.e., from 5000 MW to 7200 MW, we can see that the four curves are relatively stable and have a low fluctuation. The impact of link capacity has no obvious impact on the load balance algorithm as well. This result shows that a link capacity of 5000 MW is the starting point where the optimization algorithm begins to stabilize.

6 Conclusion

In this paper, we developed the distributed energy routing protocols *Global Optimal Energy Routing Protocol (GOER)* and *Local Optimal Energy Routing Protocol (LOER)*, which can best utilize distributed energy resources and minimize energy transmission overhead. GOER optimizes the network globally, while LOER performs suboptimal resource assignment by partitioning the grid and applying the global optimization scheme to each region. Using extensive simulation experiments, our data show that our routing protocols can more efficiently allocate energy than conventional load balancing protocols.

References

1. National Institute of Standards and Technology(NIST): NIST and the Smart Grid, http://www.nist.gov/smartgrid/nistandsmartgrid.cfm
2. U.S.Deartment of Energy: Smart Grid System Report. Technical report (2009)
3. NSF Workshop: New Research Directions for Future Cyber-Physical Energy Systems. Baltimore, MD (2009)
4. Molderink, A., Bakker, V., Bosman, M.G.C., Hurink, J.L., Smit, G.J.M.: Management and Control of Domestic Smart Grid Technology. IEEE Transactions on Smart Grid 1(2) (2010)
5. Baghaie, M., Moeller, S., Krishnamachari, B.: Energy Routing on the Future Grid: A Stochastic Network Optimization Approach. In: Power System Technology (POWERCON), pp. 24–28 (2010)
6. Conejo, A.J., Morales, J.M., Baringo, L.: Real-Time Demand Response Model. IEEE Transactions on Smart Grid 1(3) (2010)
7. Guan, X.H., Xu, Z.B., Jia, Q.S.: Energy-Efficient Buildings Facilitated by Microgrid. IEEE Transactions on Smart Grid 1(3) (2010)
8. Li, F.X., Qiao, W., Sun, H.B., Wan, H., Wang, J.H., Xia, Y., Xu, Z., Zhang, P.: Smart Transmission Grid: Vision and Framework. IEEE Transactions on Smart Grid 1(2) (2010)
9. Jin, T.D., Mechehoul, M.: Ordering Electricity via Internet and its Potentials for Smart Grid Systems. IEEE Transactions on Smart Grid 1(3) (2010)
10. Medina, J., Muller, N., Roytelman, I.: Demand Response and Distribution Grid Operations: Opportunities and Challenges. IEEE Transactions on Smart Grid 1(2) (2010)
11. Nguyen, P.H., Kling, W.L., Georgiadis, G., Papatriantafilou, M.: Distributed Routing Algorithms to Manage Power Flow in Agent-based Active Distribution Network. In: 2010 IEEE PES Innovative Smart Grid Technologies Conference Europe (ISGT Europe), pp. 11–13 (2010)
12. Lin, J., Yu, W., Yang, X.Y., Xu, G.B., Zhao, W.: On False Data Injection Attacks against Distributed Energy Routing in Smart Grid. In: ACM/IEEE Third International Conference on Cyber-Physical Systems (2012)
13. Russell, B.D., Benner, C.L.: Intelligent Systems for Improved Reliability and Failure Diagnosis in Distributed Systems. IEEE Transactions on Smart Grid 1(1) (2010)
14. Heyeck, M., Wilcox, E.R.: Interstate Electric Transmission: Enabler for Clean Energy. In: American Electric Power(AEP) Technical report (2008)
15. Kirkpatrick, S., Gelatt, C., Vecchi, M.P.: Optimal by Simulated Annealing. Science 220(4598), 671–680 (1983)
16. Nocedal, J., Wright, S.J.: Numerical Optimization. Springer, Heidelberg (1999)
17. Whitley, D.: A Genetic Algorithm Tutorial. Statistics and Computing 4, 65–85 (1993)
18. The Office of Electricity Delivery and Energy Reliability (OE): Smart Grid, http://www.oe.energy.gov/smartgrid.htm
19. U.S. Energy Information Administration: State Electricity Profiles 2009. Technical report (2011)

Performance Improvements Using Application Hints on a Multicore Embedded System

Yoondeok Ju and Moonju Park*

Abstract. Multicore processors are increasingly adopted to embedded systems like smartphones and tablets as user applications on such devices become more complex and require high performance. However, it is in doubt that the user applications for embedded systems with multi-processing capability exploit the power of the multicore CPU fully. Unlike servers or desktop PCs, power-performance balance is most important in embedded systems. Thus if an application is not carefully designed to efficiently use the multicore CPU, or the system is not aware of it, the use of multicore might result in unexpected failure, such as little performance improvement with high power consumption. In this paper, we present a framework for efficient use of multicore CPU in embedded systems. The proposed framework monitors the usage of the computing resources such as CPU cores, memory, network, and the number of threads. Then it manages the number of CPU cores to be assigned to the application using the resource usage hints. We have tested the framework using SunSpider benchmark with FireFox and Midori Web browsers on an embedded system with Exynos4412 quad-core. Experimental results show that by managing the core assignment and frequency scaling, we can improve the energy efficiency along with the performance.

Keywords: Multi-core, Embedded system, Task management.

1 Introduction

A multicore processor is a chip multiprocessor with symmetric multiple processing units (cores). Due to the increasing demands on higher performance applications in devices such as smartphones, multicore processors are now dominant in "smart"

Yoondeok Ju · Moonju Park
School of Computer Science and Engineering,
Incheon National University, Incheon, Korea
e-mail: dbsejr21@gmail.com, mpark@incheon.ac.kr

* Corresponding author.

R. Lee (Ed.): *SNPD*, SCI 492, pp. 161–172.
DOI: 10.1007/978-3-319-00738-0_12

embedded devices. Multicore processing in embedded devices is different from traditional high-performance computing; while performance is the most important issue in traditional high-performance computing, power is the first-order constraint for embedded systems [1]. Thus balancing power with performance is an important issue to exploiting the computing power of multicore processors in embedded systems.

Because the power is the first concern in embedded systems, studies on embedded systems have focused on reducing power consumption. DVFS (Dynamic Voltage and Frequency Scaling) [2] is a well-known established technology for CPU power handling. We can reduce the power consumption of CPUs by lowering the voltage level and the operating frequency when the system utilization is low. Contemporary systems support DVFS in operating system level as default. For example, Linux systems support five different kinds of CPU frequency governors that manage the voltage and frequency of CPUs according to pre-defined policies. DVFS has been studied much in various fields. Its principle and the survey of the DVFS technologies for real-time systems are surveyed in [3,4].

To efficiently use the computing power of multicore processors, system-level supports are indispensable as well as parallel programming support like OpenMP, pThreads, or MPI. System-level support techniques of tuning for application performance on multicores can be found [5,6]. Raman et al. proposed APIs to use run-time information to dynamically optimize the parallelism options [7]. Chen and John [8] proposed a program scheduling method by matching the resource demands by the programs and the core configurations. Lim et al. reported in [9] user waiting time in Linux on an embedded system could be reduced by enhancing the load-balance scheme of Linux. Lively et al. studied energy consumption and performance characteristics of applications implemented with OpenMP and MPI on a many core system in [10]. Ogura and Midorikawa showed using characteristics of applications such as CPU, memory utilization, and I/O characterization in parallel environment can improve the performance of the application [11] in cloud systems.

Though there have been lots of research results to boost application performance on multicore systems, few applications have been made to utilize the research results. Especially in embedded systems, it is hard to optimize an application for a target device because there are so many kinds of devices. For example, most smartphones get their applications from some kinds of APP market, and the applications are not optimized for the device; sometimes, an application may not even be multi-threaded. Therefore, as well as it is important to make a new application optimized for performance, it is also important to support applications which are not optimized, with improved performance and less power consumption.

In this paper, we propose a framework for efficiently supporting legacy applications on multicore embedded systems. In this framework, we monitor the execution of the application of interest using a resource monitor we developed. The resource monitor reports the characteristics of the application such as CPU utilization, memory usage, network usage, and the maximum number of concurrently running threads. Based on this information of application, a task manager determines the number of cores to be used and allocates CPU core(s) to the application. We have implemented the proposed framework onto Ubuntu Linux on an embedded board with an Exynos4412 multicore processor. We have measured the performance and the power consumption on the real target. The results using the proposed method show that the performance and the energy consumption are both improved with Linux's ondemand governor, or the performance is further improved with comparable energy consumption with Linux's conservative governor.

This paper is organized as follows. First, Section 2 describes the resource monitor we developed to find the characteristics of the applications. In Section 3, we present test results with the resource monitor applied to exemplary applications. In Section 4, we suggest a task management method using the characteristics of the applications, and present how the task management technique improves the performance and reduce the power consumption. Section 5 concludes our work.

2 Resource Monitor

To identify the performance and the power consumption characteristics of an application, we developed a resource monitor that reports how the resources including CPU cores, NIC, and memory are used by the application. The resource monitor is implemented on Linux (Ubuntu distribution). It monitors the application's usage of the following resources:

(1) CPU core utilization: /proc file system is supported by Linux maintains each thread's CPU time and CPU core on which it executed during the last jiffy. The resource monitor calculates the application's utilization per core using the information.
(2) Memory usage: It is calculated using information in /proc/[pid]/statm every 10ms ([pid] is the process ID of the application). Shared memory usage is calculated by the size of the share memory divided by the number of processes sharing the memory.
(3) Network usage: Because it is too costly to identify packets from/to the application, we estimate the network usage by accumulating the packet transmission/receive amount over jiffies when the application is running.
(4) Concurrent threads: The maximum number of threads that execute simultaneously in a jiffy.

The main difference of the developed resource monitor from already existing utilization monitors like top or htop is that our resource monitor monitors the usage of resources by ONLY the application we want to measure, while top or htop reports the resource utilization of the whole system. Our resource monitor also reports the maximum number of threads running simultaneously, which indicates how the application parallelizes its execution.

The resource monitor was implemented on Linaro Ubuntu for ARM 12.11, kernel version 3.6.10. Fig. 1 shows the user interface of the resource monitor. One can choose the application to be tested by navigating the file system. Before starting the application, one can define how much time to be tested and the sampling interval (≥10ms). Pressing the 'Execute' button starts the application and to monitor it. If you want to monitor the resource usage of an already executing application, it can be done by providing the application's process ID and pressing 'Monitor' button.

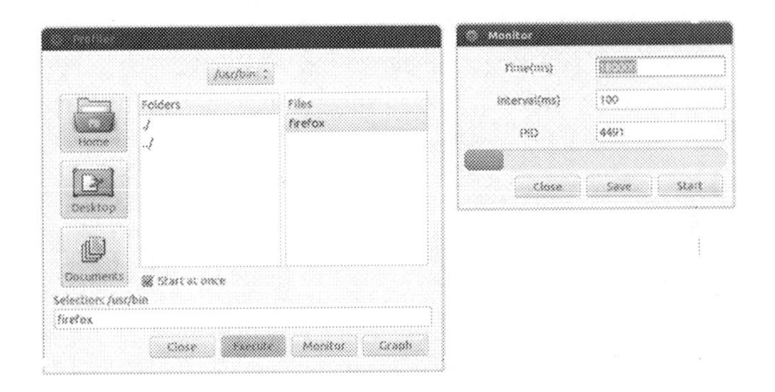

Fig. 1 Execution of an application from the resource monitor

After monitoring is done, the resource usage data are stored in a file. To avoid accessing the file system while monitoring, the usage data are maintained in the main memory during the monitoring process. Thus the monitoring time is limited by the sized of memory that can be used by the resource monitor.

3 Detecting Application Characteristics with Resource Monitor

3.1 *Test Environments*

We used Hardkernel's Odroid-X embedded board for test. Odroid-X is equipped with Samsung Exynos4412 processor, which has 4 ARM Cortex-A9 cores of 1.4GHz, 1GB DRAM, 100Mbps Ethernet port. 8GB SDHC is used for the storage media. Linaro Ubuntu for ARM 12.11 (kernel version 3.6.10) was used as its operating system. Fig. 2 shows the Odroid-X embedded board used for our experiments.

Fig. 2 Odroid-X embedded board used for tests

Power consumption of the embedded board was measured using HPM-300A power meter. The sampling interval of the power meter is 250ms, and the error is less than ±0.4%. The sampling interval of the resource monitor was 100ms. For the energy management, CPUFreq subsystem of the Linux was used with auto hot-plugging.

DVFS policy in Linux is determined by the governor of the CPUFreq subsystem. There are five governors in CPUFreq subsystem; performance, powersave, userspace, ondemand, and conservative governor. Performance governor sets CPUs run at the highest frequency, while powersave governor lets CPUs run at the lowest frequency. Userspace governor allows users to set the frequency of the CPUs. Ondemand governor is the default one, which dynamically changes the CPU frequency. It raises the frequency to the highest level when the CPU utilization exceeds the threshold. If the governor finds the utilization to be less than the threshold, it steps down the frequency to the next lower level. Conservative governor is similar to the ondemand governor, but it gradually steps up or down the frequency if the processor utilization is above or below the threshold. Thus, Linux provides two DVFS policy in effect, ondemand and conservative governor, unless users provide one by themselves. In our experiments, we measured performance and power consumption of both cases when ondemand or conservative governor is employed.

We have tested Firefox 18.0.1 and Midori 0.4.3 Web browsers, since the Web browsers are most frequently used applications on smart phones. The benchmark used in test is SunSpider JavaScript benchmark [12] version 0.9.1. The benchmark tests how fast the browser executes JavaScript codes including generating tag clouds from JSON inputs, 3D ray-tracer, encrypting strings, and uncompressing tests.

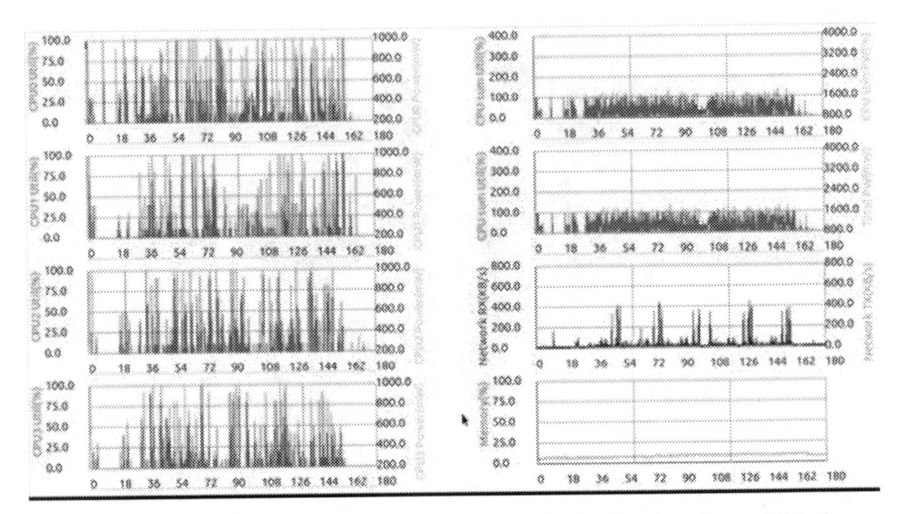

Fig. 3 Result screen of the resource monitor for the SunSpider benchmark test of Firefox

3.2 *Energy Consumption and Performance with the Original System*

Fig. 3 is the graph generated by the resource monitor showing resource usages of FireFox browser (ondemand governor is employed in this case). The resource monitor shows the per-core CPU utilization, network usage, and memory usage. We tested the Web browsers using the SunSpider benchmark three times.

The maximum number of concurrent running threads was 4 for FireFox and 2 for Midori. First, we measure the performance and the power consumption using the default ondemand governor. Note that CPU utilization is affected by which governor is employed since the speed of the CPU is changed according to the employed governor's policy. The maximum CPU utilization using ondemand governor was 140% for FireFox and 110% for Midori over 4 cores, while the average CPU utilization was 30.7% for FireFox and 28.7% for Midori over 4 cores. Therefore, it is sufficient to serve Firefox or Midori Web browser with only one core on the average, or two cores at maximum, though they used all 4 cores during the test.

Second, with conservative governor, the maximum CPU utilization was 140% for FireFox and 120% for Midori over 4 cores, while the average CPU utilization was 22.9% for FireFox and 20.8% for Midori over 4 cores.

As shown in Fig. 3, even though only one or two cores are sufficient to serve the application, the application uses all 4 cores. We guessed that it is due to Linux's load balancing, so we checked the number of migration from a core to another during the test. We assume there was a migration if a thread which had run on core A in the previous jiffy turns out to run on different core B in the last jiffy. The numbers of core migration were 553, 515, and 531 respectively during each test. As we can see from the numbers, the core migration affects the performance of the Web browser; the higher the number of migration, the longer the test time. In fact, most of the migrations of the application between cores were not necessary because the average CPU utilization was below 100%. But without such information on application behavior, the Linux load balancer moves the same application core to core, which leads to performance degradation and more power consumption.

Energy used for the SunSpider benchmark test and the performance results measured by the test are summarized in Table 1. Overall, FireFox showed better performance than Midori. Ondemand governor consumes more energy than conservative governor as expected, since it raises the CPU frequency to the top level when CPU load is high, while conservative governor gradually steps up the frequency.

Table 1 Energy Consumption & Performance of SunSpider Benchmark Test

	FireFox		Midori	
Governor	Energy(J)	Performance (ms)	Energy(J)	Performance (ms)
Ondemand	860.4	1878.3	867.6	2816.8
Conservative	815.7	2045.3	798.7	2574.7

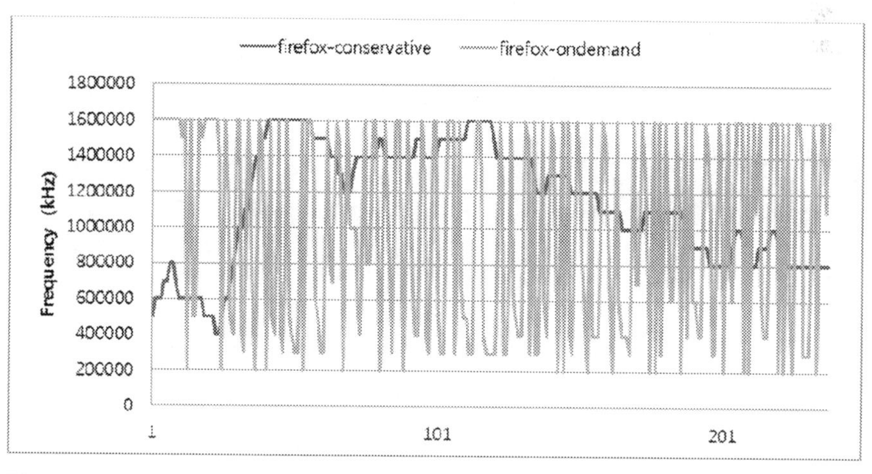

Fig. 4 Frequency changes during the first 6 seconds of the SunSpider benchmark test of FireFox (with 4 cores)

Midori with conservative governor shows better performance with less energy than with ondemand governor, while FireFox performs better with ondemad governor. The performance is related with the DVFS policy. Fig. 4 shows how CPU frequency is changed by each governor during the first 6 seconds of the benchmark test with FireFox browser. As we can see, conservative governor changes the frequency more gradually than ondemand governor. Ondemand governor raises the CPU frequency early, but it also changes the frequency even lower than conservative governor. So, the average frequency level of the application could be higher with conservative governor than ondemand governor. The average frequency level during the benchmark test of FireFox was about 810MHz with conservative governor and about 877MHz with ondemand governor, and for Midori, the frequency level was about 676MHz with conservative governor and about 671MHz with ondemand governor. So it is expected that FireFox runs faster with ondemand governor, while Midori runs faster with conservative governor.

4 Task Management Using Application Hints

With the monitoring data gathered by the resource monitor, an applications hit is generated and stored in the file system. Then the task manager allocates resources to the application using the application hint. Currently, only CPU core management is implemented.

The task manager allocates CPU cores to the application as follows. Let the maximum total CPU utilization be $Umax$ and the average total CPU utilization be $Uavg$, which lies between 0 and the number of cores. At first, the task manager assigns only $\lceil Uavg \rceil$ cores to the application and enables $\lceil Umax \rceil$ cores for the application at most. The application can use $\lceil Uavg \rceil$ always, but additional cores are enabled only when it is necessary. Task migration of the application is allowed only between the assigned cores. Unassigned cores are unplugged using clock-gating function supported by the CPU hardware.

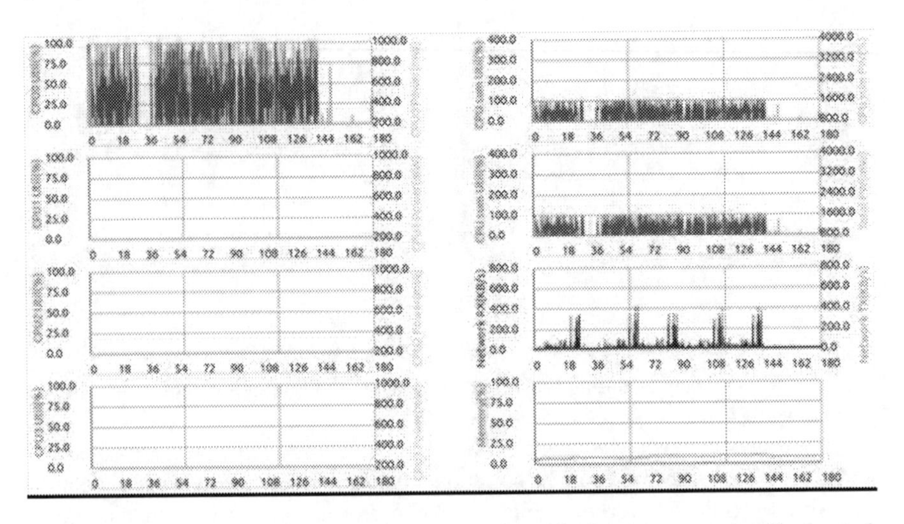

Fig. 5 Result screen of the resource monitor when Firefox is executed with the task manager

Fig. 5 shows the resource usage of FireFox executing SunSpider benchmark with the task manager. The threads of FireFox run on one core, CPU0, so the overheads of migrating threads are eliminated. It should be noticed that FireFox did not use more than one core indeed, because the task manager did not allow migration if the total utilization of FireFox's threads is below 100%. It turns out that the maximum utilization in the original Linux reached 140% because the cores run at too low speed.

Table 2 SunSpider Benchmark Test Results with Application Hints

	FireFox		Midori	
Governor	Energy(J)	Performance (ms)	Energy(J)	Performance (ms)
Ondemand	838.6 (97.6%)	1710.5 (91.1%)	850.4 (98.0%)	2453.5 (87.1%)
Conservative	821.2 (100.7%)	1686.1 (82.4%)	800.5 (100.2%)	2026.0 (78.7%)

Table 2 shows the results using the application hint and the task manager for SunSpider benchmark on FireFox and Midori. The numbers in parentheses indicate the relative percentage to the original Linux. Since the average total utilization of FireFox or Midori was below 100%, they were assigned only one core during most of their execution time, which made other cores unplugged most of the time.

Table 3 compares the energy efficiency of the applications using action (energy multiplied by time) taken during the benchmark test. Smaller number indicates higher energy efficiency (same work with less Joule-second).

Table 3 Watt·S2 of applications with SunSpider benchmark test

	FireFox		Midori	
Governor	Original	Managed with hints	Original	Managed with hints
Ondemand	1616.1	1434.4	2443.9	2086.5
Conservative	1668.4	1384.6	2056.4	1621.8

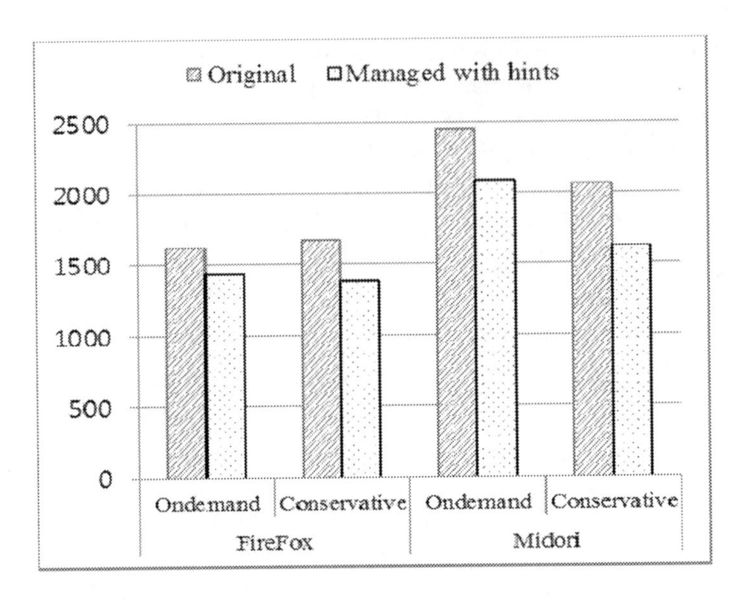

With the application hints, the task manager enhances the energy efficiency by 11.2% with ondemand governor and 17.0% with conservative governor for FireFox, and by 14.6% with ondemand governor and 21.1% with conservative governor for Midori. When the default ondemand governor was employed, the energy consumption was reduced for both applications (2.4% reduction for FireFox and 2.0% for Midori), and the benchmark performance was also improved (8.9% improvement for FireFox and 12.9% for Midori).

On the other hand, when conservative governor was employed, the energy consumption was increased for both applications (0.7% increase for FireFox and 0.2% for Midori), but we achieved more improvements in benchmark performance (17.6% improvement for FireFox and 21.3% for Midori). So with our task manager, the applications run faster with less energy consumption with ondemand governor, or run even faster with slightly more energy consumption.

5 Conclusions

In this paper, we presented a frame work for exploiting the computing power of multicore CPUs in embedded systems. The proposed framework consists of a resource monitor and a task manger; the resource monitor gathers the resource usage information of user applications, and the task manager manages task execution by assigning CPU cores to the task based on the resource usage hints. We have implemented the framework on Ubuntu Linux running on Exynos4412 quad-core ARM processor. We tested Web browser applications, FireFox and Midori, to see the effectiveness of the proposed framework. Experimental results show that the current implementations of FireFox and Midori do not exploit the multicore processor enough, both in energy consumption and performance points of view.

By restricting them to use only one core instead of four cores, the performance of the applications improved by from 8.9% to 21.3% when they were tested using SunSpider benchmark. Multiple cores degraded the performance due to the system overhead of load balancing, which would be never expected by users with the powerful multicore CPUs in their devices.

Our experiments showed that embedded systems should be aware of the application characteristics to enhance the performance and to reduce the energy consumption. DVFS policy also affects the performance and the power consumption of the applications. Our work suggests that DVFS policy for multicore embedded systems should be considered with core assignment policy, and its efficiency is related with the application's characteristics also. In our experiments, for example, FireFox showed better performance with ondemand governor than conservative governor on the original Linux, while the opposite was true for Midori. With our task manager, conservative governor was better for both browsers. From this work, we can guess there may be a better combination of DVFS policy and core assignment for performance or power consumption, and it will be our future work.

Acknowledgment. This work was supported by Basic Science Research Program through the National Research Foundation of Korea (NRF) funded by the Ministry of Education, Science and Technology (Grant No. 2012R1A1A1010948).

References

1. Munir, A., Ranka, S., Grodon-Ross, A.: High-Performance Energy-Efficient Multicore Embedded Computing. IEEE Transactions on Parallel and Distributed Computing Systems 23(4), 684–700 (2012)
2. Weiser, M., Welch, B., Demer, A.J., Shenker, S.: Scheduling for Reduced CPU Energy. In: Proceedings of the 1st USENIX Conf. on Operating Systems Design and Implementation, pp. 13–23 (1994)
3. Nasro, M.-A., Wang, Y., Xing, J., Nisar, W., Kazmi, A.: Towards Dynamic Voltage Scaling in Real-Time Systems - A Survey. International Journal of Computer Sciences and Engineering Systems 1(2), 93–103 (2007)
4. Chen, J.-J., Kuo, C.-F.: Energy-Efficient Scheduling for Real-Time Systems on Dynamic Voltage Scaling (DVS) Platforms. In: Proceedings of IEEE International Conference on Embedded and Real-Time Computing Systems and Applications, pp. 28–38 (2007)
5. Karcher, T., Pankratius, V.: Auto-Tuning Multicore Applications at Run-Time with a Cooperative Tuner. Technical Report, 2011-4, Karlsruhe Institute of Technology, Germany (2011)
6. Zwinkau, A., Pankratius, V.: AutoTunium: An Evolutionary Tuner for General-Purpose Multicore Applications. In: Proceedings of 2012 IEEE 18th Conference on Parallel and Distributed Systems, pp. 392–399 (2012)
7. Raman, A., Kim, H., Oh, T., Lee, J.W., August, D.I.: Parallelism Orchestration using DoPE: the Degree of Parallelism Executive. In: Proceedings of the 32nd ACM SIGPLAN Conference on Programming Language Design and Implementation, pp. 26–37 (2011)

8. Chen, J., John, L.K.: Efficient Program Scheduling for Heterogeneous Multi-core Processors. In: Proceedings of the 46th ACM/IEEE Design Automation Conference, pp. 927–930 (2009)
9. Lim, G., Min, C., Eom, Y.: Load-Balancing for Improving User Responsiveness on Multicore Embedded Systems. In: 2012 Linux Symposium (July 2012)
10. Lively, C., Wu, X., Taylor, V., Moore, S., Chang, H.-C., Cameron, K.: Energy and Performance Characteristics of Different Parallel Implementations of Scientific Applications on Multicore Systems. The International Journal of High Performance Computing Applications 25(3), 342–350 (2011)
11. Ogura, D.R., Midorikawa, E.T.: A Methodology for Characterizing Applications for Cloud Computing Environments. In: Proceedings of International Conference on Parallel and Distributed Computing and Systems, pp. 78–85 (2012)
12. http://www.webkit.org/perf/sunspider/sunspider.html

Seasonal Rainfall Prediction Using the Matrix Decomposition Method

Hideo Hirose, Junaida Binti Sulaiman, and Masakazu Tokunaga

Abstract. The matrix decomposition is one of the most powerful methods in recommendation systems. In the recommendation system, even if evaluation values in a matrix where users and items are corresponding to row and column are provided incompletely, we can predict the vacant elements of the matrix using the observed values. This method is applied to a variety of the fields, e.g., for movie recommendations, music recommendations, book recommendations, etc. In this paper, we apply the matrix decomposition method to predict the amount of seasonal rainfalls. Applying the method to the case of Indian rainfall data from 1871 to 2011, we have found that the early detection and prediction for the extreme-value of the monthly rainfall can be attained. Using the newly introduced accuracy evaluation criterion, risky, we can see that the matrix decomposition method using cylinder-type matrix provides the comparative accuracy to the artificial neural network result which has been conventionally used.

1 Introduction

Flood events are becoming more frequent and intense in many countries around the world. One of the major concerns in the world is the recent increase of catastrophic flood situations. Many researches point out that in the coming decades the situation may become worse due to the climate change. Accurate forecasting of rainfall has been one of the most important issues in hydrological research, because early warnings of severe weather, made possible by timely and accurate forecasting can help prevent casualties and damages caused by natural disasters. The intricacy of the atmospheric processes that generate rainfall makes the physical modeling of rainfall overly parameterize. A possible solution to this is to

Hideo Hirose · Junaida Binti Sulaiman · Masakazu Tokunaga
Kyushu Institute of Technology, Fukuoka, 820-8502 Japan
e-mail: hirose@ces.kyutech.ac.jp

R. Lee (Ed.): *SNPD*, SCI 492, pp. 173–185.
DOI: 10.1007/978-3-319-00738-0_13

construct the rainfall forecasting system based on the simple data structure. In this paper, we propose to use the recommendation system to predict the seasonal rainfall amount.

The matrix decomposition is one of the most powerful methods in recommendation systems. In the recommendation system, even if evaluation values in a matrix where users and items are corresponding to row and column are provided incompletely, we can predict the vacant elements of the matrix using the observed values. In this paper, we apply the matrix decomposition method to predict the amount of seasonal rainfalls, the real Indian rainfall data from 1871 to 2011 [8].

The matrix decomposition method is a kind of nonparametric approach to estimate the vacant elements in a incomplete matrix. From a viewpoint of parametric approach, there may be another approach to estimate them. Although one of the methods using the item response theory can be seen in [12, 18], we describe the nonparametric approach here.

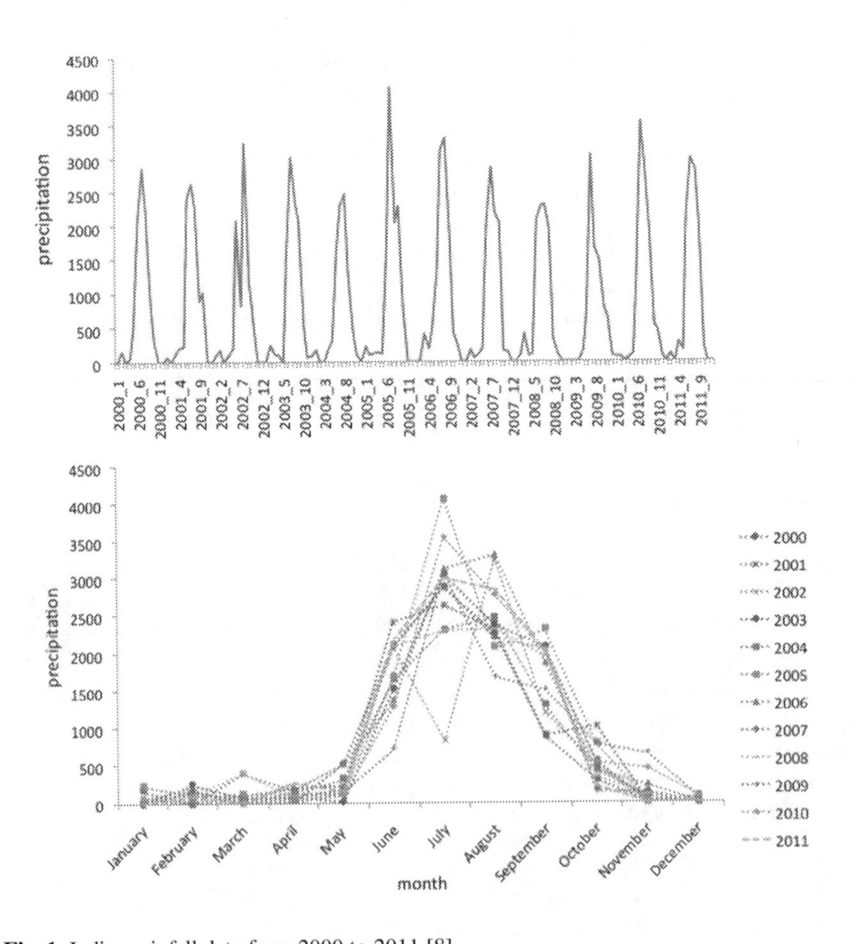

Fig. 1 Indian rainfall data from 2000 to 2011 [8]

2 Traditional Rainfall Forecasting Approaches

Generally, there are two major approaches to forecast rainfall, conceptual (physical) modeling and system theoretical modeling [5, 9]. Conceptual (physical) approaches are designed to make close to the physical mechanisms which govern the hydrologic processes (simplified as physical laws), and usually based on the characteristics and knowledge of a specific catchment (watershed). Conceptual approaches are widely applied in hydrological forecasting, and particularly for non-stationary hydrological data and continuous variables data tracking required [7, 24]. This approach for rainfall forecasting may not be feasible due to significant calibration data (explanatory variables) to reasoning rainfall is not easily to be collected [25], and the volume of rainfall calculations require sophisticated mathematical tools [6].

System theoretical approaches are employed mapping models to formulate the relationships between inputs and outputs without consideration of the physical structure processes. Developed by Box and Jenkins [3], the autoregressive moving average with exogenous inputs (ARMAX) models have been one of the most popular approaches in time series forecasting. However, a fundamental limitation for time-series forecasting models is their inability to predict changes that are not based on the past data, particularly for the nonlinearity of rainfall related variables.

Artificial neural networks (ANN) are kind of a neuron based model which is able to capture nonlinear relationship between input variables and output. The basic neural networks which consist of input-output processes come in a form of general equation,

$$Y = F(X) + \varepsilon \tag{1}$$

where the Y is the model output, X is a k-dimensional input vector whose components are denoted by $X_n (n = 1, \ldots, k)$ and ε represents a Gaussian randomness. The processing power of ANN lies on the hidden layer where the learning are taken place. Selection of hidden layer nodes has become one of the most important issues in ANN architecture. Small number of hidden nodes may cause difficulties to learn data for the network, whereas large number of nodes could make the network memorize too in detail for learning and generalization, which might lead to the problem of local minima [15]. However, properly selected of hidden nodes can help the network adjust to larger fluctuation of target function and allow to consider the presence of volatility which is normally found in meteorological type of data. Here, we adopted a suggestion by Fletcher and Goss [15] in determining the range of hidden nodes which will give good results.

ANN has been extensively applied in regression problems such as rainfall forecasting [9], [15], groundwater prediction [4] and hydrology forecasting [1]. The popularity of ANN is due to its ability to learn and generalize from examples even when the data contain errors or are incomplete [9]. The structure of ANN regression model for time series can be expressed as

$$X(t+1) = g(X(t), X(t-1), \ldots, X(t-n), u_1(t), u_2(t), \ldots, u_i(t)) + e(t), \tag{2}$$

where $X(t)$ represents a vector of output at time t, $g()$ represents a non-linear function approximates by ANN, $X(t-i)$, $(i=1,\ldots,n)$ are the past values of the time series itself and u_i are the exogenous variables at time t. $e(t)$ is a mapping error to be minimized by some cost function. The applications of ANN to rainfall forecasting are seen also in [19, 20, 21, 22]

3 Matrix Decomposition Method

The problem description for the recommendation system by using the matrix decomposition is simple [23]. When we obtain a matrix in which observed evaluation values by users and items are incompletely occupied, then we can predict the vacant elements of the matrix using the observed values. That is, the unobserved value for (i,j) element in the matrix is estimated (to $\hat{x}(i,j)$) by using the observed values of $x(i,j)$; see Figure 2.

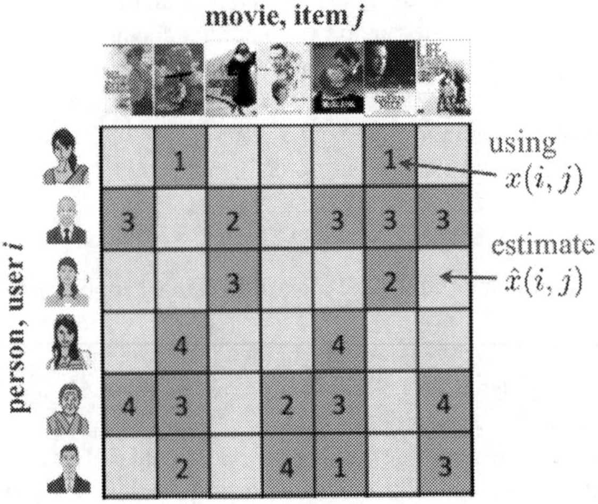

Fig. 2 The idea for the recommendation system using the matrix decomposition

A variety of methods have been proposed to solve this kind of problem. The use of the similarities such as the correlation coefficient or cosine is the primary method to estimate the vacant elements; k-nearest neighbor approach using the similarity is the next step; a promising methodology is the use of matrix decomposition (matrix factorization) [11]. Although the winner of the Netflix competition [10] used the ensemble technique combining all the known methods together [2], one of the most effective and major methods is still the matrix decomposition.

The singular-value decomposition, abbreviated as SVD, is one of the factorization algorithms for various applications which include computing the pseudo-inverse, least squares fitting of data, matrix approximation, and determining the rank, range and null space of a matrix. Suppose $P \in R^{m\times n}$, $U \in R^{f\times m}$, and $M \in R^{f\times n}$ are matrices. A simple idea that a matrix factorization $P = U^T M$ produces the missing data of score matrix V leads us to the use of the collaborative filtering. Thus, the matrix decomposition, which is also used for recommendation systems (see references [17, 16, 26]), is used for the least square method here. That is, we want to find the matrix U and M by minimizing the target function E such that sum of the squares of the difference between the observed score $V(i,j)$ and the predicted score $P(U_i,M_j)$,

$$E = \frac{1}{2}\sum_{i=1}^{m}\sum_{j=1}^{n} I(i,j)(V(i,j) - P(U_i,M_j))^2, \tag{3}$$

where $P(U_i,M_j)$ denotes the (i,j) element of $U^T M$. This idea of the matrix decomposition is derived by the usual SVD formulation such that $A = U\Sigma V^*$ where U and V are orthonormal and Σ provides the singular values in the diagonal elements. If Σ is absorbed by either or both U and V, we can accomplish the matrix decomposition of A.

Suppose $V \in R^{m\times n}$ is the score matrix of m users and n items, and $I \in 0,1^{m\times n}$ is its indicator. The SVD algorithm finds two matrices U and M as the feature matrix of users and items. That is, each user or item has an f-dimension feature vector and f is called the dimension of the SVD. A prediction function p is used to predict the values in V. The value of a score $V(i,j)$ is estimated by $p(U_i,M_j)$, where U_i and M_j represent the feature vector of user i and item j, respectively. Once U and M are found, the missing scores in V can be predicted by the prediction function.

For stable and robust computing, the optimization of U and M is actually performed by minimizing the sum of squared errors between the existing scores and their prediction values with penalty factors:

$$\begin{aligned} E = {} & \frac{1}{2}\sum_{i=1}^{m}\sum_{j=1}^{n} I(i,j)(V(i,j) - p(U_i,M_j))^2 \\ & + \frac{k_u}{2}\sum_{i=1}^{m} \| U_i \|^2 + \frac{k_m}{2}\sum_{j=1}^{n} \| M_j \|^2, \end{aligned} \tag{4}$$

where k_u and k_m are regularization coefficients to prevent overfitting; $\|\cdot\|$ means the Frobenius norm (l^2 norm). This formulation is a kind of the ridge regressions. The most common prediction function is the dot product of feature vectors. That is, $p(U,M) = U^T M$. The optimization of U and M thus becomes a matrix factorization problem where $V \approx U^T M$.

When using the prediction function of $p(U,M) = U^T M$, the objective function and its negative gradients have the following forms:

$$\begin{aligned} -\frac{\partial E}{\partial U_i} = {} & \sum_{j=1}^{n} I(i,j)\left((V(i,j) - p(U_i,M_j))\frac{\partial p(U_i,M_j)}{\partial U_i} \right) \\ & - k_u U_i, \end{aligned}$$

$$-\frac{\partial E}{\partial M_j} = \sum_{i=1}^{m} I(i,j)\left((V(i,j) - p(U_i, M_j))\frac{\partial p(U_i, M_j)}{\partial M_j}\right) \\ - k_m M_j. \tag{5}$$

One can then perform the optimization of U and M by the descent gradient method or stochastic descent gradient method by using the algorithm,

$$U^{(t+1)} \leftarrow U^{(t)} + \mu \frac{\partial E}{\partial U}, \\ M^{(t+1)} \leftarrow M^{(t)} + \mu \frac{\partial E}{\partial M}, \tag{6}$$

where μ is the learning rate.

4 Rainfall Prediction Using the Matrix Decomposition Method

The seasonal trend of the amount of the rainfall to each year can be described in a matrix; monthly for column and year for row. For example in the case of the Indian rainfall data from 1871 to 2011 [8], we can collect data yearly from January in 1871 to December 2011 as shown in Figure 3 which includes the data in Figure 1. The number of monthly data in a year is 12. This incomplete matrix is used to the matrix decomposition, and we may predict the unobserved elements in the matrix.

Why we had an idea to apply the matrix decomposition method to the rainfall data can be explained by a similarity to the seasonal infectious disease prediction which was successful [13, 14].

	January	February	March	...	October	November	December
1871	303	47	6.75	...	172	100	19
...	...	...	...	...	...	...	...
2008	13	101	410	...	338	103	13
2009	28	0	32	...	786	651	82
2010	72	69	32	...	540	448	83
2011	0	96	13	...	207	8	4

Fig. 3 The matrix data of the Indian rainfall data from 1871 to 2011

In Figure 4, on the top left, we show the predicted result for February to December using the yearly data from 1871 to 2010 and January data. Just below it, we show the case for prediction of March to December. Similarly, eleven cases are shown each in the figure. Figure 5 shows exactly the same eleven cases altogether. We can see that an extreme-value for the amount of the rainfall can be well predicted. Here, we have used the parameters such that $\mu = 5 \times 10^{-6}$, and $k_m = k_u = 2 \times 10^{-5}$, and $f = 10$.

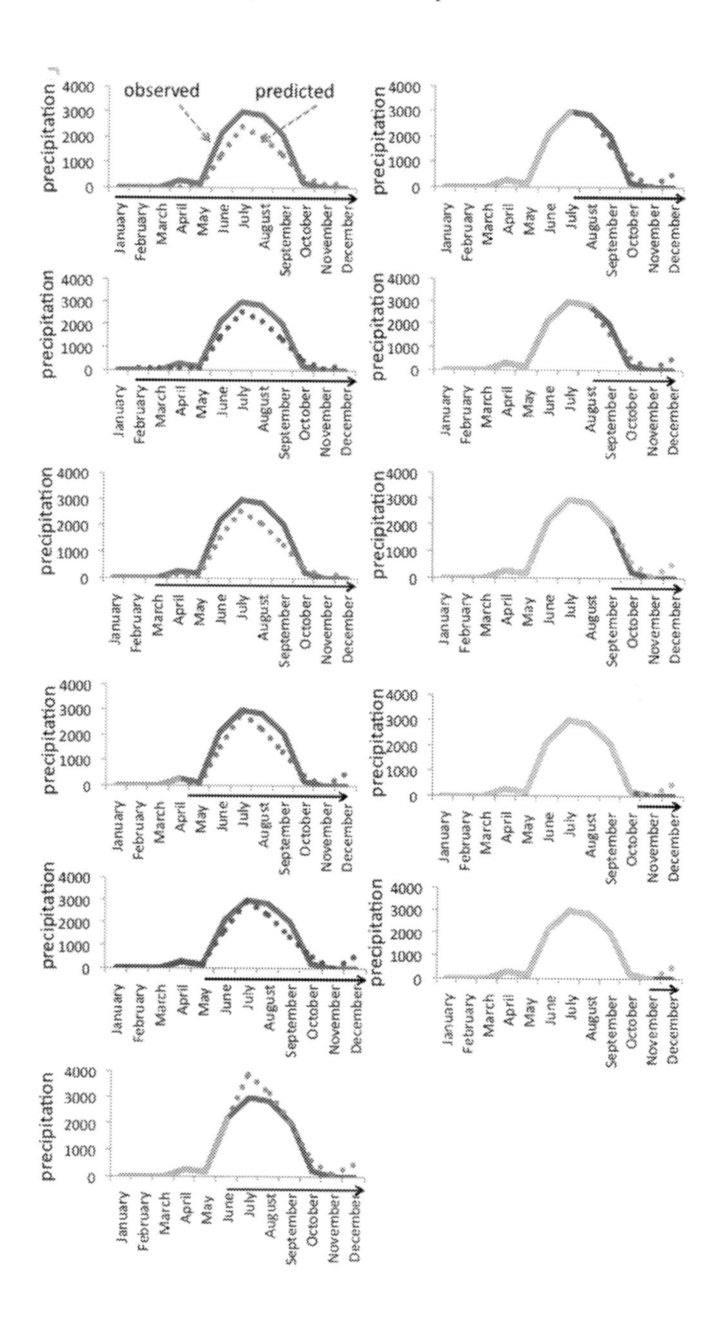

Fig. 4 The predicted result for 2011 using the yearly data from 1871 to 2010 using the matrix decomposition method: each prediction to December

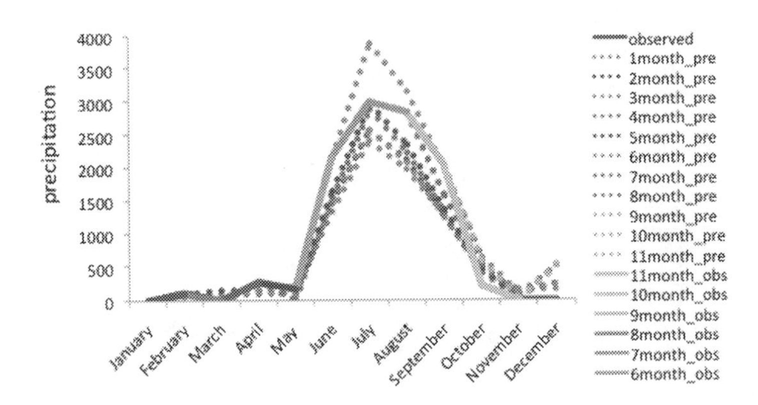

Fig. 5 The predicted result for 2011 using the yearly data from 1871 to 2010 using the matrix decomposition method: eleven cases predictions to December

Figure 6 shows a similar figure to Figure 5; In Figure 6, the artificial neural network method is used for prediction.

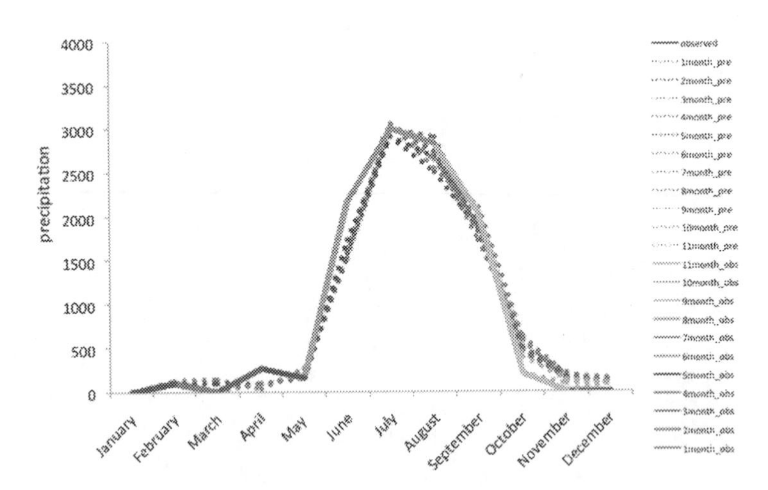

Fig. 6 The predicted result for 2011 using the yearly data from 1871 to 2010 using the artificial neural networks method: 11 predictions to December

5 Rainfall Prediction Using the Cylinder-Type Matrix

The methodology mentioned above does not consider the continuity of the yearly data such as December to next year January. To overcome this inconvenience, we have to use the cylinder-type matrix as shown in Figure 7.

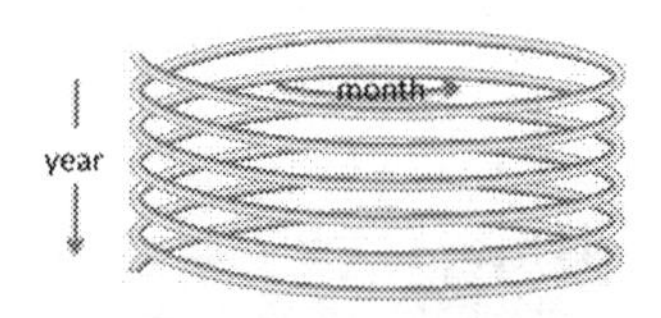

	January	February	March	...	October	November	December	January
1871	303	47	6.75	...	172	100	19	
...	...	...	...	...	...	...	...	
2008	13	101	410	...	338	103	13	
2009	28	0	32	...	786	651	82	
2010	72	69	32	...	540	448	83	
2011	0	96	13	...	207	8	4	

Fig. 7 Cylinder-type matrix

In this paper, we modified this kind of matrix by adding next year January data to the right-side of the matrix. This modified matrix is called the quasi-cylinder-type matrix. Applying the matrix decomposition method in the quasi-cylinder-type matrix to the Indian rainfall data from 1871 to 2011, we have found that the early detection and prediction for the rainfall can be obtained; see Figure 8.

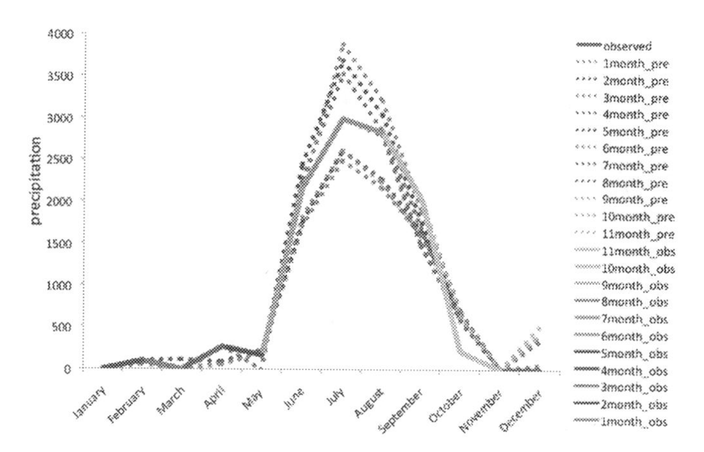

Fig. 8 The predicted result for 2011 using the yearly data from 1871 to 2010 using the quasi-cylinder-type matrix

6 Accuracy of the Prediction

6.1 Root Mean Squared Error, RMSE

To check if the proposed method provides a good accuracy or not, we have compared the *RMSE* (root mean squared error) obtained by the matrix decomposition methods with that by using the other method, ANN method in time series analysis for comparison.

The formula of ANN for Indian rainfall data is expressed as

$$y_t = f(x_{(t-1)}, x_{(t-2)}, \cdots, x_{(t-n)}), \tag{7}$$

where $y_{(t)}$ is one-step ahead forecast output and $x_{(t-n)}$, $(n = 13)$ are the past values of rainfall. The architecture of ANN is 13-5-1 as in form of input, hidden and output nodes (I-H-O). The inputs are 13 months of earlier rainfall values while the output is one month forecast. The network is trained with Levenberg-Marquart algorithm and used the tansig (hyperbolic tangent sigmoid transfer function) and pureline activation functions in hidden and output layers.

The *RMSE* is defined by

$$RMSE = \sqrt{\frac{1}{|T|}\sum_{i,j} I(i,j)(\hat{x}(i,j) - x(i,j))^2},$$
$$(|T| = \sum_{i,j} I(i,j),\ I(i,j) \text{ for test data}). \tag{8}$$

We will use the data before 2010 as training, and data of 2011 as test. However, in the case of the matrix decomposition method, we have to include at least one

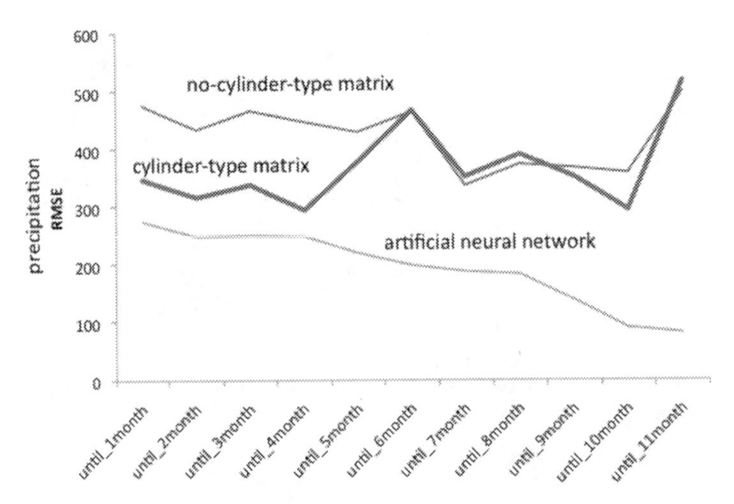

Fig. 9 The *RMSE* for accuracy of prediction by 1) matrix decomposition method without cylinder-type matrix, 2) matrix decomposition method with cylinder-type matrix, and 3) artificial neural networks

observation to 2011 row in the matrix. The cases of adding 1 month, 1-2 months, ... , 1-11 months data are considered here. The *RMSE* computed in such a manner are shown in Figure 9.

This does not show the superiority of the proposed method over the conventional methods using ANN. However, contrary to the conventional method, the proposed method provides the higher amount of the rainfall from earlier, which could be preferable in a risk viewpoint.

6.2 Risk Based Weighted L_1 Norm, Risky

We define a new accuracy evaluation criterion here from a risk viewpoint. If the predicted value is under-estimated, the cost for compensation could be high. On the contrary, if the over-estimated prediction is obtained, we may disregard the cost for preparing to the disaster. Thus, we can attain this by the use of weighting to L_1 norm error, such that:

$$\begin{aligned} &risky = \frac{1}{|T|}\sum_{i,j} w(i,j)I(i,j)|\hat{x}(i,j) - x(i,j)|, \\ &if\ \hat{x}(i,j) - x(i,j) > 0,\ then\ w(i,j) = 1, \\ &otherwise,\ w(i,j) = k > 1 \\ &(|T| = \sum_{i,j} I(i,j),\ I(i,j) \text{ for test data}). \end{aligned} \tag{9}$$

Figure 10 shows the *risky* by using 1) no-cylinder-type matrix, 2) cylinder-type matrix, and 3) ANN; here, we have used the value of k as $k = 2, 4$. We can see that the matrix decomposition method using cylinder-type matrix provides the comparative accuracy to the ANN result when using the data before May.

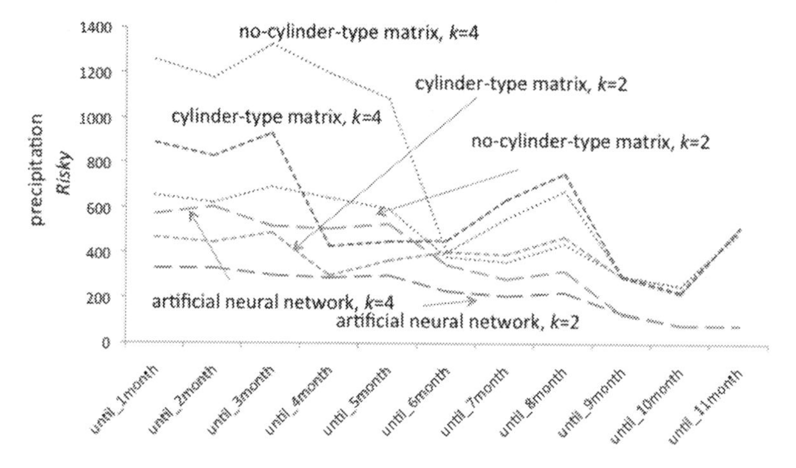

Fig. 10 *Risky* for accuracy of prediction by 1) matrix decomposition method without cylinder-type matrix, 2) matrix decomposition method with cylinder-type matrix, and 3) artificial neural networks.

7 Concluding Remarks

To predict the vacant elements of the incomplete matrix using the observed values in the matrix, the matrix decomposition is one of the most promising methods, which is often used in the recommendation systems. This method is applied to a variety of the fields, e.g., for movie recommendations, music recommendations, book recommendations, etc. In this paper, we apply the matrix decomposition method to predict the amount of seasonal rainfalls. Applying the method to the case of Indian rainfall data from 1871 to 2011, we have found that the early detection and prediction for the extreme-value of the monthly rainfall can be attained. Comparing the root mean squared error between the predicted and observed data, we have found that the proposed method does not show the superiority of the proposed method over the conventional methods using the artificial neural networks method. However, contrary to the conventional method, the proposed method provides the higher amount of the rainfall from earlier, which could be preferable in a risk viewpoint. Using the newly introduced accuracy evaluation criterion, risky, we can see that the matrix decomposition method using the cylinder-type matrix provides the comparative accuracy to the artificial neural networks method result.

Acknowledgements. The authors would like to thank to Mr. Takenori Sakumura for his cooperation.

References

1. Araghinejad, S., Azmi, M., Kholghi, M.: Application of artificial neural network ensembles in probabilistic hydrological forecasting. Journal of Hydrology 407, 94–104 (2011)
2. Bell, R.M., Bennett, J., Koren, Y., Volinsky, C.: The million dollar programming prize. IEEE Spectrum (May 2009)
3. Box, G.E.P., Jenkins, G.M.: Time Series Analysis: Forecasting and Control, Holden-Day, San Francisco (1976)
4. Dash, N.B., Panda, S.N., Remesan, R., Sahoo, N.: Hybrid neural modeling for groundwater level prediction. Neural Computing and Applications 19, 1251–1263 (2010)
5. Duan, Q.: A global optimization strategy for efficient and effective calibration of hydrologic models, Ph.D. dissertation, University of Arizona, Tucson (1991)
6. Duan, Q., Sorooshian, S., Gupta, V.K.: Optimal use of sce-ua global optimization method for calibrating watershed models. Journal of Hydrology 158, 265–284 (1994)
7. Druce, D.J.: Insights from a history of seasonal inflow forecasting with a conceptual hydrologic model. Journal of Hydrology 249, 102–112 (2001)
8. Indian data, `http://www.tropmet.res.in/staticpage.phppageid=53`
9. Luk, K.C., Ball, J.E., Sharma, A.: An application of artificial neural networks for rainfall forecasting. Mathematical and Computer Modelling 33, 683–693 (2001)
10. Netflix prize, `http://www.netflixprize.com/`
11. Netflix Update: Try This at Home,
`http://sifter.org/simon/journal/20061211.html`
12. Hirose, H., Sakumura, T.: Item response prediction for incomplete response matrix using the EM-type item response theory with application to adaptive online ability evaluation

system. In: IEEE International Conference on Teaching, Assessment, and Learning for Engineering 2012 (TALE 2012), pp. 8–12 (2012)
13. Hirose, H.: A seasonal infectious disease spread prediction method by using the singularvalue decomposition. In: The First BMIRC International Symposium on Frontiers in Computational Systems Biology and Bioengineering (2013)
14. Hirose, H., Nakazono, T., Tokunaga, M., Sakumura, T., Sumi, S.M., Sulaiman, J.: Seasonal infectious disease spread prediction using matrix decomposition method. In: The 4th International Conference on Intelligent Systems, Modelling and Simulation (ISMS 2013), pp. 121–126 (2013)
15. Hung, N.Q., Babel, M.S., Weesakul, S., Tripathi, N.K.: An artificial neural network model for rainfall forecasting in Bangkok. Hydrology and Earth System Sciences 13, 1413–1425 (2009)
16. Salakhutdinov, R., Mnih, A.: Probabilistic matrix factorization. In: Proc. Advances in Neural Information Processing. Systems 20 (NIPS 2007), pp. 1257–1264. ACM Press (2008)
17. Paterek, A.: Improving regularized singular value decomposition for collaborative filtering. In: Proceedings of KDD Cup and Workshop (2007)
18. Sakumura, T., Kuwahata, T., Hirose, H.: An adaptive online ability evaluation system using the item response theory. In: Education and e-Learning (EeL 2011), 51–54 (2011)
19. Sulaiman, J., Hirose, H.: A Method to predict heavy precipitation using the artificial neural networks with an application. In: 7th International Conference on Computing and ConvergenceTechnology (ICCIT 2012), pp. 687–691 (2012)
20. Sumi, S.M., Zaman, M.F., Hirose, H.: A neural network ensemble incorporated with dynamic variable selection For rainfall forecast. In: International Conference on Software Engineering, Artificial Intelligence, Networking and Parallel/Distributed Computing (SNPD 2011), pp. 7–12 (2011)
21. Sumi, S.M., Zaman, M. F., Hirose, H.: A novel hybrid forecast model with weighted forecast combination with application to daily rainfall forecast of fukuoka city. In: Nguyen, N.T., Kim, C.-G., Janiak, A. (eds.) ACIIDS 2011, Part II. LNCS, vol. 6592, pp. 262–271. Springer, Heidelberg (2011)
22. Sumi, S.M., Zaman, M.F., Hirose, H.: A rainfall forecasting method using machine learning models and its application to Fukuoka city case. International Journal of Applied Mathematics and Computer Science 22, 841–854 (2012)
23. Takimoto, S., Hirose, H.: Recommendation systems and their preference prediction algorithms in a large-scale database. Information 12, 1165–1182 (2009)
24. Vieux, B.E., Cui, Z., Gaur, A.: Evaluation of a physics-based distributed hydrologic model for flood forecasting. Journal of Hydrology 298, 155–177 (2004)
25. Yapo, P., Gupta, V.K., Sorooshian, S.: Automatic calibration of conceptual rainfall-runoff models: sensitivity to calibration data. Journal of Hydrology 181, 23–48 (1996)
26. Zhang, S., Wang, W., Ford, J., Makedon, F., Pearlman, J.: Using singular value decomposition approximation for collaborative filtering. In: Seventh IEEE International Conference on E-Commerce Technology (CEC 2005), pp. 257–264 (2005)

Effectiveness of the Analysis Method for the Impression Evaluation Method Considering the Vagueness of Kansei

Shunsuke Akai, Teruhisa Hochin, and Hiroki Nomiya

Abstract. In recent years, Kansei becomes important. However, the conventional methods are difficult to evaluate Kansei because Kansei is vague. An impression evaluation method considering the vagueness of Kansei has been proposed. This evaluation method makes a subject evaluate impression spatially. A method for analyzing the evaluation results has been proposed and the results of analysis have been shown. This analysis method shows average values and coefficients of variation of scores of the evaluation results spatially. However, only a few subjects joined the evaluation experiment. In this paper, an evaluation experiment is newly conducted, and more evaluation results are obtained. These results are analyzed, and characteristics of the impression of objects and the dispersion among subjects could easily be obtained. It is shown that this analysis method is useful for examining characteristics of impression of objects.

Keywords: Analysis method, impression evaluation, Kansei, vagueness, impression dispersion.

1 Introduction

In recent years, in addition to functions and convenience, Kansei topics such as design have become important. Kansei is a word that means how people feel. Because Kansei is vague, it is difficult to precisely capture and quantify it. Moreover, because Kansei differs for each person, it is difficult to evaluate.

The semantic differential (SD) method [1] is often used to evaluate Kansei. Because the SD method digitizes impressions, it enables statistical processing and

Shunsuke Akai · Teruhisa Hochin · Hiroki Nomiya
Department of Information Science, Kyoto Institute of Technology, Kyoto 606-8585, Japan
e-mail: m2622001@edu.kit.ac.jp, {hochin,nomiya}@kit.ac.jp

R. Lee (Ed.): *SNPD*, SCI 492, pp. 187–201.
DOI: 10.1007/978-3-319-00738-0_14

makes it possible to perform various analyses. However, to enable statistical processing, the evaluation is required to be performed in a predefined range. As a result, it is difficult to evaluate vague aspects of Kansei. A method enabling the evaluation of the vagueness of Kansei is required. Although various research efforts concerning the expression of Kansei have been conducted [2-4], such an evaluation method has not yet been established. An impression evaluation method considering the vagueness of Kansei has been proposed in order to overcome this issue [5]. The proposed method uses a plane containing impression words. The impression of an object is specified by circling the areas matching the impression. The degree of matching of the impression is expressed by the painting color. This method is called the Impression Evaluation Method by Space (IEMS).

A method for analyzing the evaluation results through the IEMS has been proposed [6]. This analysis method shows average values and coefficients of variation of the scores of the evaluation results spatially. Although the results of analysis have been shown, only a few subjects joined the experiment [6]. Effectiveness of the proposed analysis method could not sufficiently be shown.

In this paper, an evaluation experiment is newly conducted. The number of subjects joining the experiment is sufficient. Another set of pictures, which are targets of evaluation, is used in the experiment. This experiment results in the clarification of effectiveness of the proposed method. This method could show the impression which almost all of subjects commonly feel and the one which only a few subjects feel.

This paper is organized as follows. Section 2 shows the evaluation method of Kansei called IEMS, and the impression evaluation system based on IEMS. Section 3 describes an analysis method for IEMS. An experiment is conducted in Section 4. Some considerations are made in Section 5. Finally, Section 6 concludes this paper.

2 Impression Evaluation Method by Space

2.1 Kansei Space

The IEMS uses a Kansei space. The Kansei space is the space imagined in evaluating the impression of an object. For example, when the impression of a landscape is evaluated, the impression can be expressed with words such as “beautiful.” In addition, the degree of fitness to an impression word can also be expressed. It is believed that people have in their minds some impression expression items, and they compute the degree of fitness of each item to the impression of an object.

Impression words are usually used as impression expression items because words can easily express an impression (something considered by the authors to be very important) [5]. In the Kansei space, the more similar the impressions of two impression words are, the closer these impression words are. It is thought that the Kansei space changes according to experience and learning. The Kansei space depends on the person.

2.2 Summary of IEMS

The IEMS uses a plane containing impression words as the Kansei space. The impression of an object is specified by circling the areas matching the impression. The degree of matching the impression is expressed by the painting color. In other words, the more closely the impression is matched, the darker the color that is used. The color gray is used for these areas. A special brush, which makes painted areas dark gradually, is used.

Because it is difficult to create the Kansei space mentioned above from scratch, IEMS provides to users a commonly used baseline Kansei space. Users can modify this baseline space as needed. The baseline Kansei space has been obtained by applying multi-dimensional scaling [7] to the results of a questionnaire about the dissimilarity between impression words [5]. It is shown in Fig. 1. The impression words of the baseline Kansei space are called baseline impression words.

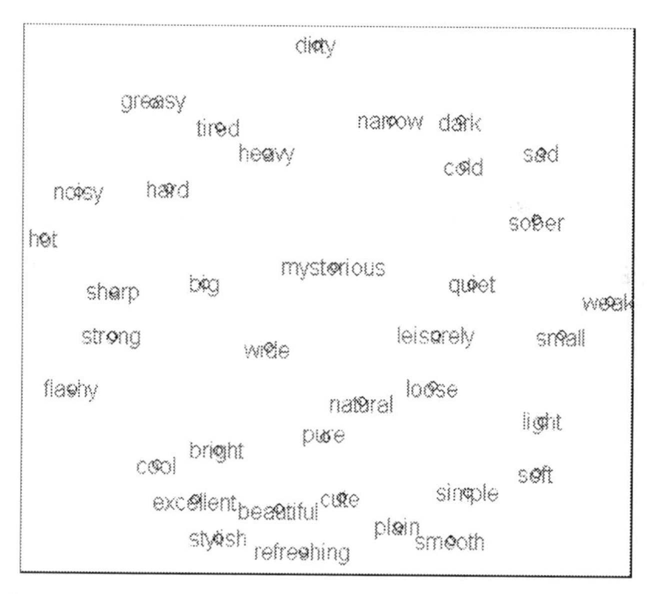

Fig. 1 Baseline Kansei space

2.3 Evaluation Rules in IEMS

The evaluation rules used in the IEMS are as follows:

1. The impression words selected to express the impression of an object are circled.
2. One or more impression words can be circled in an area.
3. One or more circled areas can exist.
4. If the impression points-of-view are different, the circled areas are different. For example, when we see a starry sky, some people feel bright and dark. They

can evaluate this impression by circling the area near the word “bright” and circling another area near the word “dark.”

5. If the space does not have the desired impression words, the impression words can be added.
6. If a user does not agree with the position of impression words, the impression words can be moved.
7. Two or more Kansei spaces can be used when users cannot sufficiently evaluate impressions through one Kansei space. In other words, when users want to evaluate different impression points-of-view by using the same impression word, or when the areas overlap, users can evaluate an impression through two or more Kansei spaces.

2.4 Evaluation Example

An evaluation example is shown in Fig. 2. It depicts that the impression of an object is beautiful, bright, and wide. In addition, the degree of darkness of a color expresses how much it matches an impression. Using this method, a user can also evaluate the impression between impression words. A vague impression can be evaluated by expressing the degree of darkness of a color. In addition, even if a matching impression word cannot be found, impression words having similar meanings can be used. Furthermore, two words with opposite meanings at the same time can also be used. Such cases cannot be evaluated using the SD method.

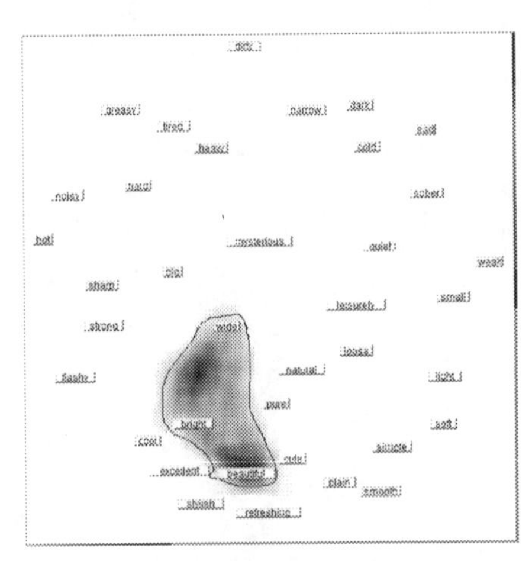

Fig. 2 Evaluation example

2.5 Impression Evaluation System

The impression evaluation system based on IEMS has been implemented [5]. This system enables subjects to evaluate their impression of a picture. The initial screen of this system is shown in Fig. 3. The Kansei space is presented in the center of the screen, while a picture is presented on the right side. Pictures are presented in a random order. A subject evaluates his or her impression of a picture in the space by using the buttons and text boxes available in the left and bottom sides of the screen.

In this system, the height and the width of the Kansei space are 600 pixels. Each pixel has 256 scales as the degree of darkness (0 shows white and 255 shows black).

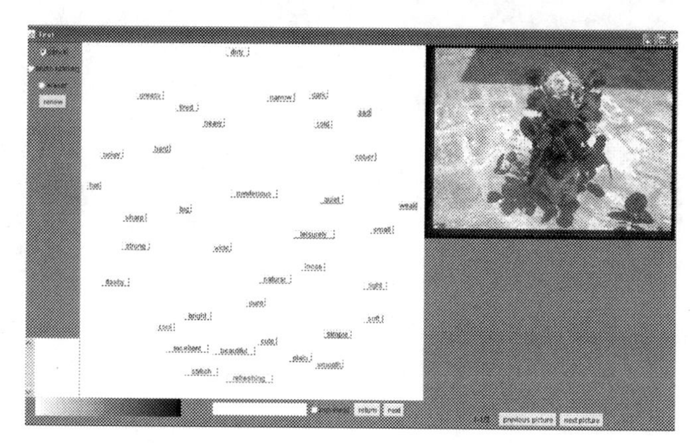

Fig. 3 The impression evaluation system

3 Analysis Method

By using IEMS, impression evaluation results are obtained from subjects. However, it is difficult to compare evaluation results because the positions of impression words are different. Here, we consider the situation that objects are evaluated by using the baseline Kansei space, where impression words are neither moved nor added. It means that Rule 5 and Rule 6 are excluded from the evaluation rules described in **2.3** during the evaluation of objects. In this situation, the objects are evaluated through the same indicator because the impression words and their positions are identical in the Kansei space used in the evaluation. In this case, the average value and the coefficient of variation of the degrees painted by subjects at each pixel could be used to obtain the tendency of the evaluation results. The coefficient of variation is the ratio of the standard deviation to the average value.

As the average values and the coefficients of variation are calculated for all of the pixels in the baseline Kansei space, these could spatially be presented. The space whose pixels represent average values is called an AVG space, and the space whose pixels represent coefficients of variation is called a CV space [6].

4 Experiment

4.1 Experimental Settings

An impression evaluation experiment with six pictures is conducted using IEMS. Six pictures used in this experiment are shown in Fig. 4. The order of presenting pictures is random. Japanese words are used in the experiment. Eleven subjects participated in this experiment. Ten subjects are males, and one is female. All subjects are university students in their twenties, majoring in Information Science.

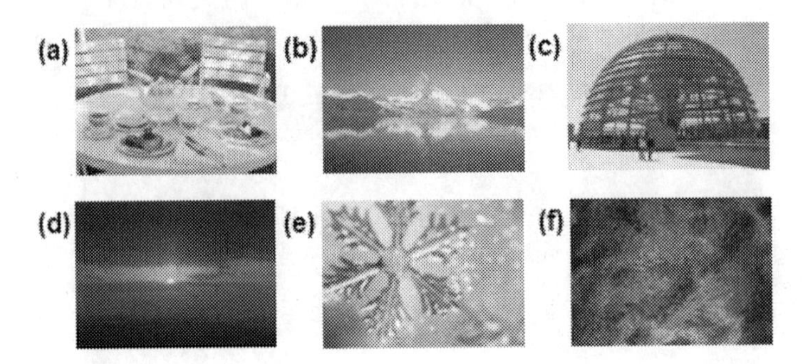

Fig. 4 Six pictures used in the experiment [8-10]

4.2 Experimental Results

Using these data, the average values and the coefficients of variation are calculated (only circled areas are calculated). If one subject evaluates impression of a picture by using two or more Kansei spaces, those evaluations are put together as his or her evaluation result on one Kansei space. Specifically, if circled areas are piled up, the pixels of these areas are averaged. In other areas, values of pixels are added.

The AVG spaces of Pictures (a) to (f) shown in Fig. 4 are shown in Fig. 5, 7, 9, 11, 13, and 15, respectively. The CV spaces are shown in Fig. 6, 8, 10, 12, 14, and 16, respectively.

For the average value, the darker the place is, the higher the average value is. It shows the average impression of the subjects. For the coefficient of variation, the darker the place is, the higher the coefficient of variation is. It shows the dispersion of impressions among the subjects.

In Fig. 6, the coefficients of variation of the place near the words "leisurely," "loose," and "quiet," and the place near the words "beautiful," "stylish," "refreshing," "cute," and "bright" are small. It is thought that these areas show the common impression of Picture (a). Furthermore, in Fig. 5, the average values of the place near the words "leisurely" and "loose," and the places near the word "stylish" and near the word "cute" are large. These tendencies are shown in Table 1. It is thought that the average impression of Picture (a) is "leisurely," "loose," "stylish," and "cute."

The similar tendencies can be found in the other pictures. The impression words around the places darkly colored in the AVG space and those lightly colored in the CV space are shown in Table 1. The places darkly (lightly, respectively) colored in the AVG (CV) space mean that their pixels have large average values (small coefficients of variation).

Table 1 Impression words with large average values and those with small coefficients of variation

Picture	Large average value	Small coefficient of variation
(a)	[leisurely, loose] [stylish] [cute]	[leisurely, loose, quiet] [beautiful, stylish, refreshing, cute, bright]
(b)	[quiet, leisurely, loose, natural] [big] [beautiful]	[quiet, leisurely, loose, natural, pure] [big, wide] [bright, cool, excellent, beautiful, refreshing]
(c)	[big] [hard] [heavy]	[big] [hard] [heavy]
(d)	[beautiful] [natural] [quiet]	[beautiful, excellent, cool, bright] [natural] [quiet]
(e)	[cold] [natural, pure] [beautiful]	[cold] [natural, pure] [beautiful, excellent, stylish, cool, plain]
(f)	[mysterious] [tired]	[mysterious] [tired, greasy, heavy, dirty] [flashy, strong] [noisy, hot] [dark] [bright]

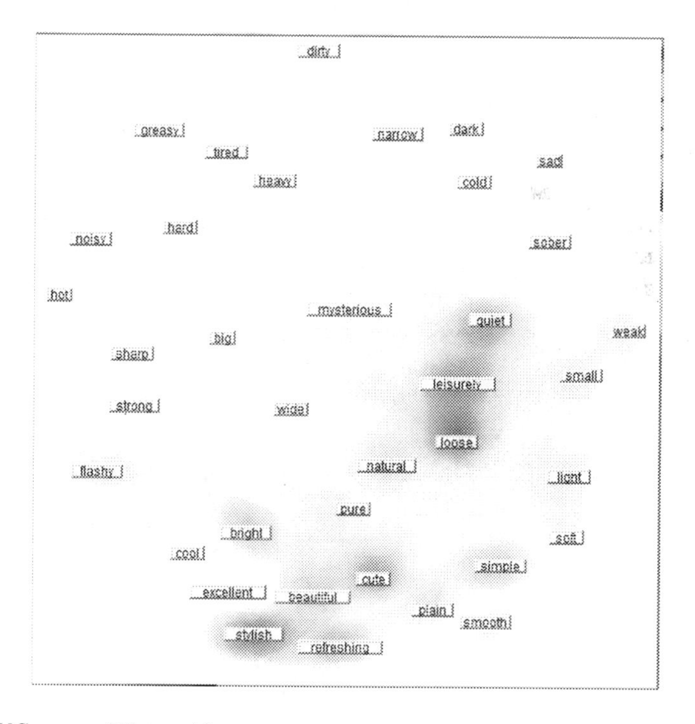

Fig. 5 AVG space of Picture (a)

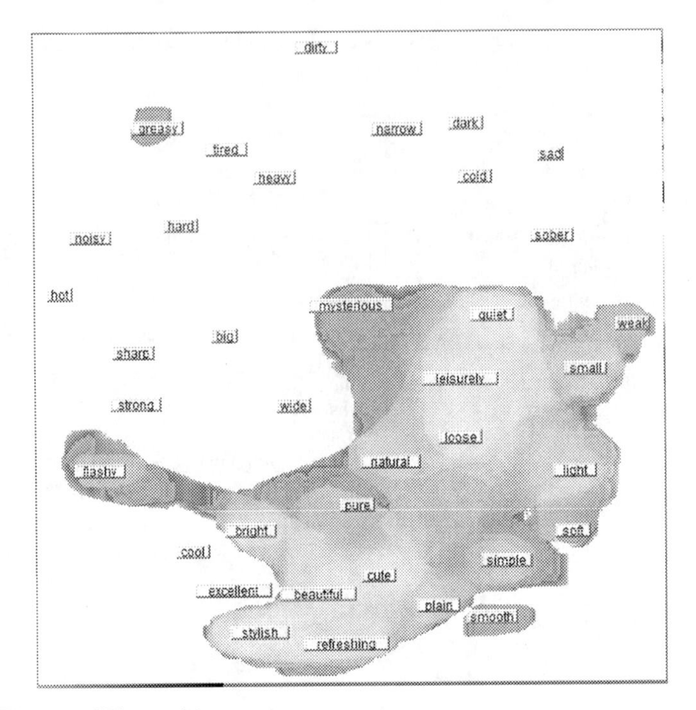

Fig. 6 CV space of Picture (a)

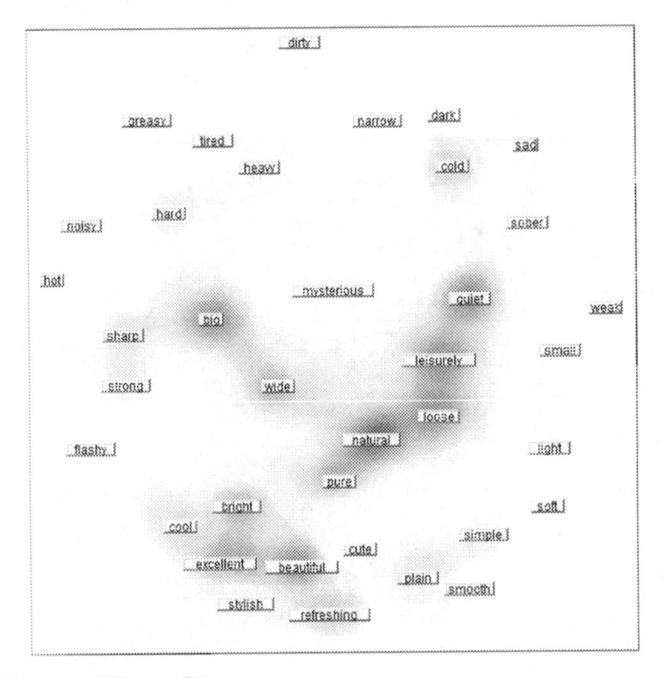

Fig. 7 AVG space of Picture (b)

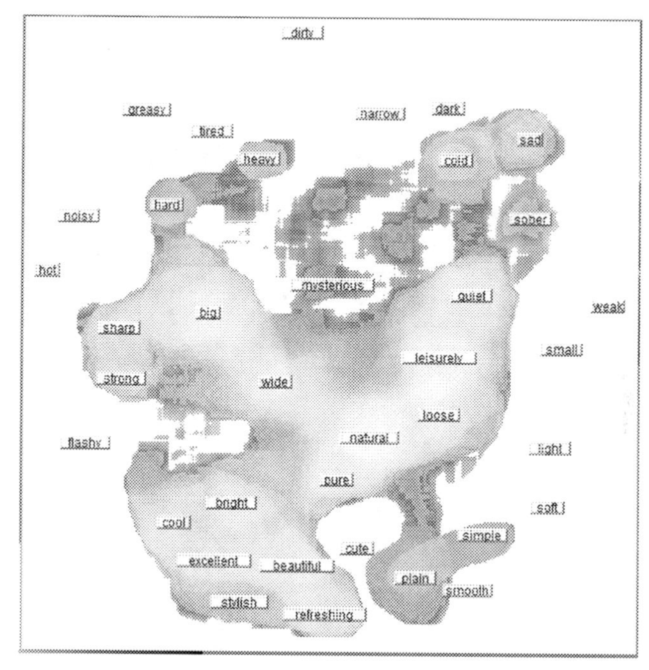

Fig. 8 CV space of Picture (b)

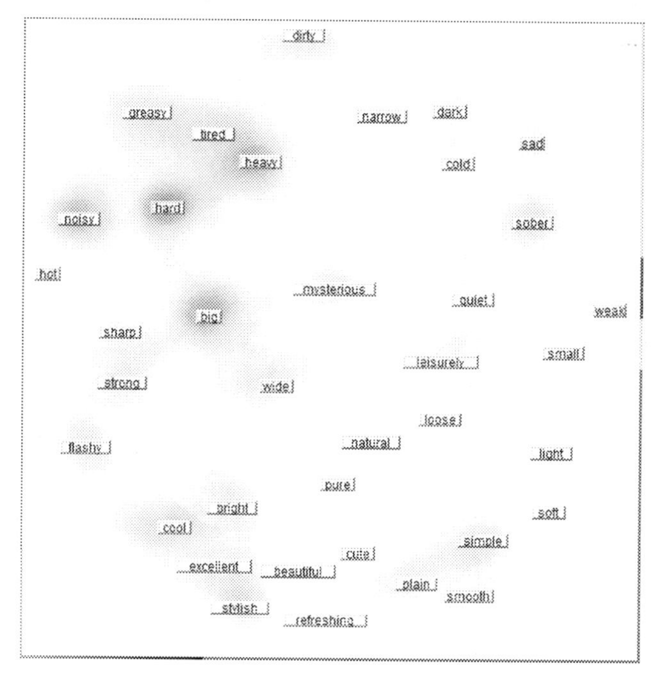

Fig. 9 AVG space of Picture (c)

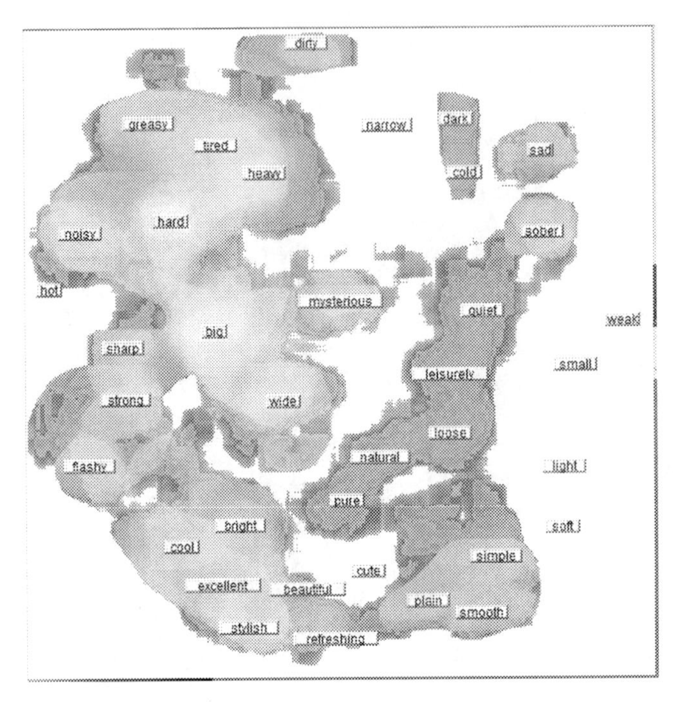

Fig. 10 CV space of Picture (c)

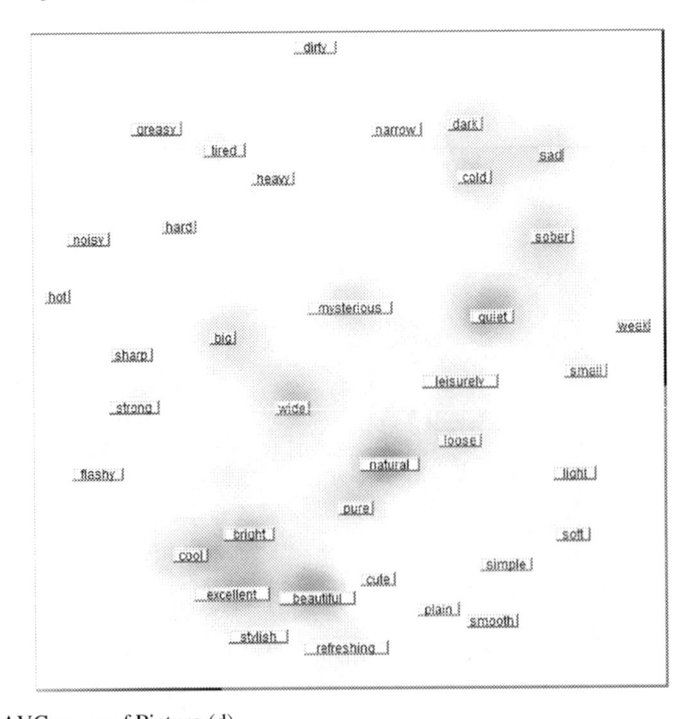

Fig. 11 AVG space of Picture (d)

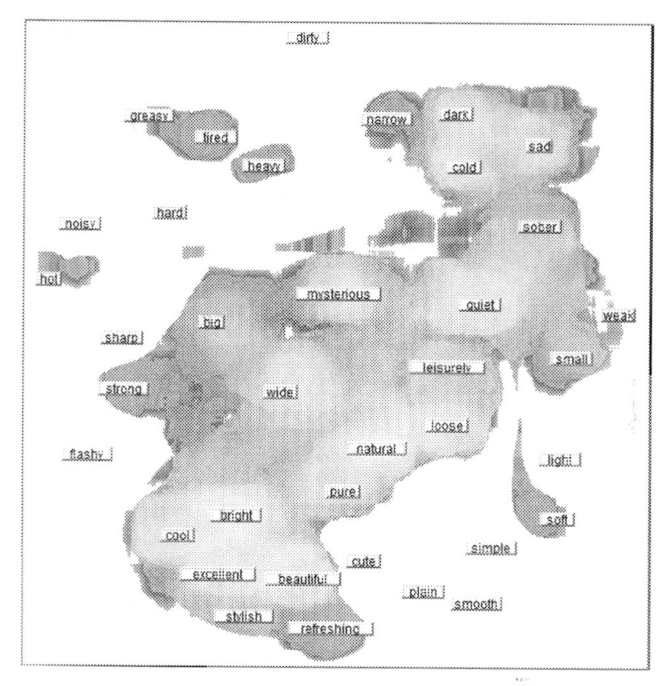

Fig. 12 CV space of Picture (d)

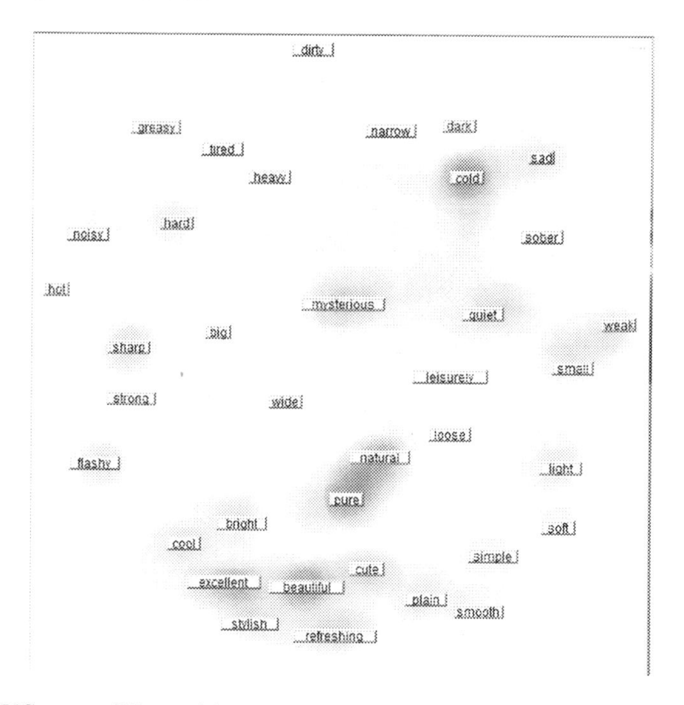

Fig. 13 AVG space of Picture (e)

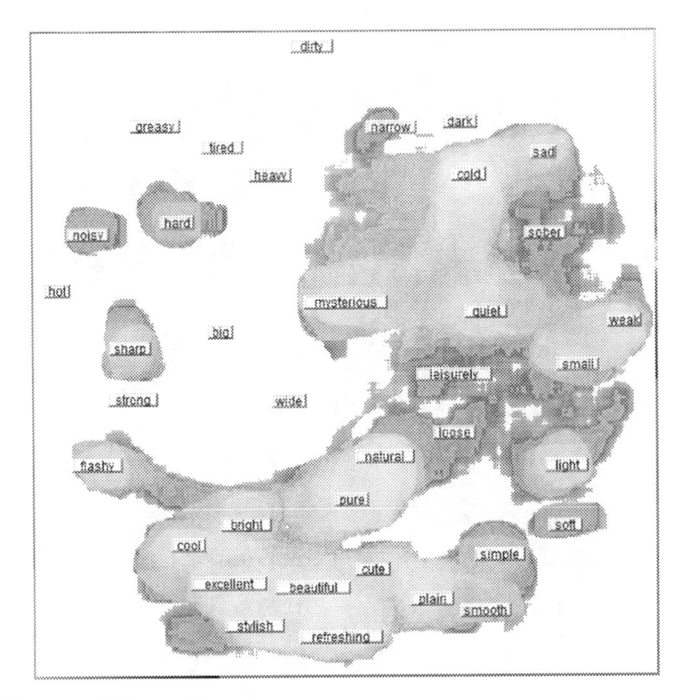

Fig. 14 CV space of Picture (e)

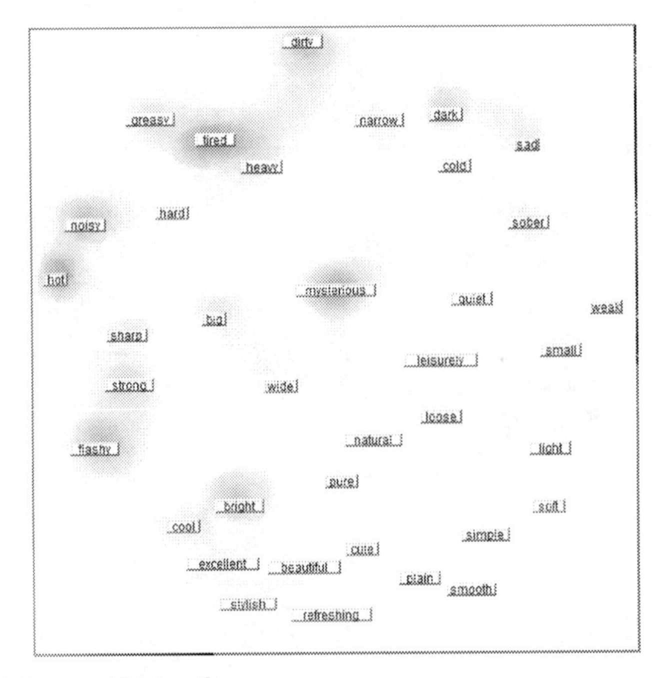

Fig. 15 AVG space of Picture (f)

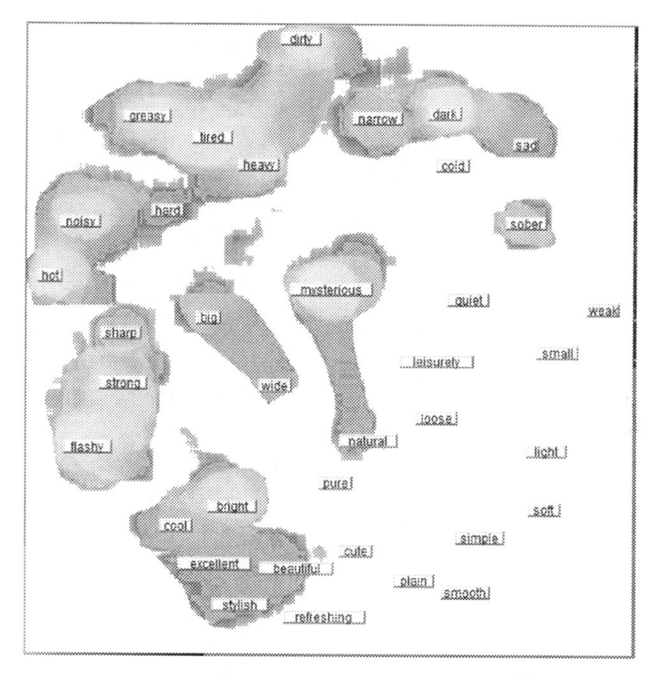

Fig. 16 CV space of Picture (f)

5 Considerations

The area, where the average values of degrees of pixels are big and their coefficients of variation are small, shows the impression that the subjects commonly feel. These areas appear for all of the pictures.

In figures of coefficient of variation, there are the area with dark color and the one with light color. The former indicates the uncommon impression because dark color means a large coefficient of variation. The latter indicates the common impression. By using the proposed analysis method, the dispersion of impression among subjects can easily be understood.

In Fig. 11 and 15, the places near the word "bright" and near the word "dark" was painted. In other words, the impressions of Pictures (d) and (f) are not only "bright" but also "dark." It is impossible to evaluate opposite impressions received at the same time through the SD method, while it is possible through the IEMS.

In Fig. 5, the average values of the place between "leisurely" and "loose" are big. This means that the subjects feel the impression between "leisurely" and "loose." Similar phenomenon could be found in Fig. 13. The place between "natural" and "pure" is darkly painted. The subjects feel the impression between "natural" and "pure." We would like to emphasize that the impression the subjects feel is the one between "natural" and "pure" rather than the impression of "natural" and that of "pure." It is said that IEMS could successfully derive the subject's feeling. This may be valuable for designer.

The proposed analysis method is said to be useful for examining characteristics of impression of objects.

In Fig. 6 and Fig. 16, the colored areas are smaller than the other CV spaces. It is thought that many subjects feel similar impression to Pictures (a) and (f). On the other hand, in Fig. 8, 10, 12, and 14, the colored areas are large. It is thought that subjects feel various impressions to Pictures (b), (c), (d) and (e). Categorizing the subjects may give valuable information to us. By adapting the type of subject, we might produce user-oriented products, services and so on.

A set of pictures different from the one used in the previous experiment [6] was used in the experiment in this paper. Although a different set of pictures was used, the same ability of the analysis method could be shown as in the previous experiment. Therefore, the ability of the analysis method is considered to be independent of the pictures used in the experiments. Moreover, many subjects joined the experiment. It is also shown that the characteristics, which can be derived by using the proposed analysis method, do not depend on a specific person in a limited number of subjects.

6 Conclusion

This paper newly conducted an evaluation experiment, and more evaluation results were obtained. By calculating the average value and the coefficient of variation of each pixel, the average impression of a picture and the dispersion among subjects were obtained. The characteristics of the impression of an object can easily be obtained from these data because the average values and the coefficients of variation are spatially shown. It is shown that the proposed analysis method is useful for examining characteristics of impression of object.

Conventional methods do not consider the dispersion among subjects. However, because subjects feel various impressions, the dispersion among subjects should be considered. The proposed analysis method is useful for considering the dispersion among subjects.

The proposed analysis method can be used only in a certain condition. That is, its usage is limited to the evaluation through the baseline Kansei space. It could not be used for the evaluation results through the Kansei space where impression words are moved and/or added. Establishing the analysis method, which could be applied to the evaluation results through such a Kansei space, is in future work. Categorizing subjects is considered to be valuable. The categorization of subjects is also included in future work.

References

[1] Osgood, C.E., Suci, G.J., Tannenbaum, P.H.: The Measurement of Meaning. University of Illinois Press (1957)

[2] Choi, H., Okazaki, A.: Development of the KANSEI evaluation program for a product - Focusing on visualization of the conceptual model. In: Proc. of 13th Annual Conference of Japan Society of Kansei Engineering (2011) (in Japanese)

[3] Tazaki, S., Okazaki, A.: A method to quantify KANSEI directly. In: Proc. of 10th Annual Conference of JSKE (2008) (in Japanese)
[4] Kashiwazaki, N.: Proposition of KANSEI-parameters to show amount of KANSEI changing by outside influence. Journal of Japan Society of Kansei Engineering 4(1), 31–34 (2004) (in Japanese)
[5] Akai, S., Hochin, T., Nomiya, H.: Impression Evaluation Method Considering the Vagueness of Kansei. In: Proc. of 13th ACIS International Conference on Software Engineering, Artificial Intelligence, Networking and Parallel/Distributed Computing (SNPD 2012), pp. 385–390 (2012)
[6] Akai, S., Hochin, T., Nomiya, H.: A method for analyzing the results obtained through the impression evaluation method considering the vagueness of Kansei. In: Proc. of the 1st International Symposium on Affective Engineering (ISAE 2013), pp. 27–32 (2013)
[7] Schiffman, S.S., Reynolds, M.L., Young, F.W.: Introduction to multidimensional scaling - Theory, Methods, and Applications. Academic Press (1981)
[8] Moonpocket: SOZAIJITEN 102, Datacraft (2001)
[9] Moonpocket: SOZAIJITEN 186, Datacraft (2007)
[10] Moonpocket: SOZAIJITEN 177, Datacraft (2006)

Representation, Analysis and Processing of Student Counseling Cases

Naotaka Oda, Aya Nishimura, Takuya Seko, Atsuko Mutoh, and Nobuhiro Inuzuka

Abstract. Student counseling becomes an important roles in University. Many colleges have been paying effort for the task. The field of student counseling has not introduced computational technique. The paper challenges a new approached to contribute the field by giving a representation of counseling records. By giving a representation effective analysis and processing of cases are possible. First, we give a formal representation of counseling cases based on our observation that persons and their network and changes of the network are important structure of cases. Second, we try to capture characteristics of cases by giving attributes and transitions of relations that compose networks. Third, a similarity measure is defined for cases based on the formal representation and attributes. These proposal are examined by counseling cases, which are prepared by rearranging real cases.

1 Introduction

Universities are expected to pay more effort to support students, which brings forward student counseling[2, 4]. The function of student counseling has started around 1950's and the earliest student counseling section in Japan was launched in the University of Tokyo in 1953. Since then the importance of the function has been recognized in wide area of the student support field.

More recently Japanese Ministry of Education, Science and Culture (which is reorganized as Ministry of Education, Culture, Sports, Science and Technology in 2003) published a report (called the Hironaka report) about fulfilling of student life, 2000. The report states that the student support is a part of education that should be given in universities and emphasizes the move from universities for faculties to universities of students. Another report, called Tomabechi report, by JASSO (Japan Student Service

Naotaka Oda · Aya Nishimura · Takuya Seko · Atsuko Mutoh · Nobuhiro Inuzuka
Nagoya Institute of Technology, Nagoya 466-8555, Japan
e-mail: {oda,nishimura,seko}@nous.nitech.ac.jp,
{mutoh,inuzuka}@nitech.ac.jp

R. Lee (Ed.): *SNPD*, SCI 492, pp. 203–217.
DOI: 10.1007/978-3-319-00738-0_15

Organization) in 2007 had an impact for this matter. It requires all university staffs to take effort to student support with help by student counseling professionals. These two reports made a strong reason to enforce universities to this direction.

Student supporting has been gathering attention because many kinds of students and many ways of studies become common and because students meet many difficulties in their academic career and consideration of their future. These movements have increased also because of recent economic depression and anxiety to aging society.

Despite of the attention to the student support, universities have budget shortfall in human resource and we need time for fostering consensus of academic staffs. One of our motivations of this research is to help the student support task using computer and AI technology. The technology may construct a human network to the task of student support and gives improvements of support systems. Our motivation is not only the reason for efficiency improvements. Introducing computer and AI technology to student support may help to understand difficulties in student life, and it helps to model human minds in real social activities and communication.

2 Student Counseling Systems

We have been developing recording systems for student counseling. The systems are expected to use in student counseling sections in universities. Since we have started this project we developed two series of systems. One is to use in student support sections. The college of the authors started to use an earlier version of the system series in 2007 and in 2011 they replaced it to a more effective version. Now the system is used by the student support section, the student career advising section and the foreign student support section of the college. Since we installed the earlier system more than 12,000 counseling records involving more than 1,800 students are recorded.

The other series of systems are not complete and are experimental. We have studied ideal functions for student support and counseling, the functions which are not only for student counseling but also for understanding students' minds. Through development of the system we have been discussing the function of recording system of student counseling. Here we point out few notes for the recording system and importance of human relationships. This is the main topic in this paper.

Through our study, we revealed that records of counseling are different from medical records in many points. We may say that the main purpose of medical records is to gather information[3]. With an electronic medical recording system all information of clients from medical examinations and past medical history are collected on a desktop of a doctor. On the other hand in counseling situations, a counselor need to remember the episode of clients and relives their experience as they feel. The counseling recording system need to keep episodes as a flow of situations according with time. The system also describes situations with feeling which clients felt with surrounding events and persons. Then we can find out that we need to give framework to describe situations with people and feeling, and flow of situations. By the formalization of episodes, we may search an episode by conditions and organize episodes by their similarity. Advanced functions for the system may be expected to give analysis and to discover to help student counseling practice. The records should be analyzed to give answers for difficulties.

In this paper we give a representation framework of student counseling records. First we describe our observation for student counseling cases and give a formal representation for records, which gives a database scheme with which we designed a student counseling system in Section 3. In Section 4 we sort out relations among persons appearing in cases by classifying them into objective relations and subjective relations. Then Section 5 demonstrates analysis using transitions of the relations. Section 6 deals with computational aspects of treatment of counseling cases. It gives a similarity measure based on the observation in Section 3 and an approximated but efficient procedure for it. Finally Section 7 concludes with some future works.

3 Representation of Student Counseling Cases

3.1 Structure of Counseling Records and Database Schema

We have observed many cases of student counseling with a cooperation of a professional counselor. Student counseling starts by a contact of a student who has a difficulty in his/her student life or by a motion of other people, such as a friend, a parent, and a supervising teacher of the student with a counselor or someone in charge of a student support desk. Then the counselor sees and talks with the consulter student on his/her chief complaint and tries to resolve the complaint. It may continue several times, dozens of times or more until his/her difficulty goes away or he/she leaves the college.

Through the observation we find the following facts about counseling cases.

- A case includes many sessions and to grasp the overview of session helps to understand the situation of the case.
- The content talked in a session does not necessarily happened at the time of the session. It may be a past time. The order of the contents does not match with the order of sessions.
- A case involves many persons surrounding the consulter student in the problem and relationships between the student and the persons talked or among persons.
- The changes of situations in a case may correspond with changes of relations among persons.

According with these observations we designed a database schema shown in Fig. 1, which includes four tables, cases, sessions, persons and relations and an ontology that manages relations. For the ontology we discuss in Section 4. Cases are registered in the tables cases and all sessions in every cases are registered in the table sessions with foreign key case ID to the table cases. Each recored of sessions has the date of session and the date of content. The table relations keeps relations appearing in sessions with session ID as the foreign key for the start date and the end date. All information about persons are managed in persons, which includes

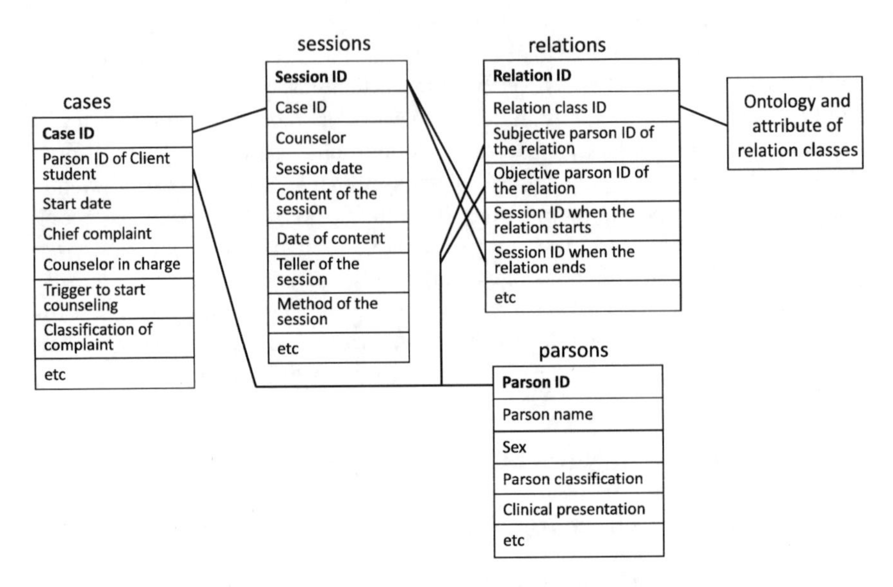

Fig. 1 Database schema for student counseling records

students who came for counseling and persons appearing in sessions. Persons appearing in sessions are connected to sessions through relations.

This schema has problems when we use this schema in the actual counseling operations. A student may talk in a session contents for more than one date, a date of content may not be clear. These problems remain for the future work but the schema is enough to use for our analysis in this paper.

3.2 Formal Structure of Cases

Formally we can treat a counseling case c as a triple (P_c, S_c, R_c), where P_c is a set of all persons appearing in the case c, S_c is a set of sessions of which c consists, and R_c is a set of relations appearing in c. The P_c, S_c and R_c for all cases correspond to the three tables, persons, sessions, and relations, respectively.

Now we consider only a case and we write the triple, (P, S, R), without index of the case. We use the same symbols for each session and its ID. That is, s may denotes a session and also its ID. Likewise we treat for a person and a relation. The representation of a person p in P includes its ID and other information, such as the name. The representation of a session s in S includes its ID, the date of the session, denoted by $talked(s)$, the date of experiencing the contents talked in the session, $date(s)$, and other information. The representation of a relation r in R includes its ID, the session of which the content corresponds with the beginning of the relation, denoted by $bgn(r) \in S$, and the session of which the content corresponds with the finish of the relation $fini(r) \in S$. A relation also includes two persons from and to whom the relation holds, denoted by $from(r) \in P$ and $to(r) \in P$, respectively, and other information.

For a session $s \in S$ there is a network among persons surrounding the consulter student. The **network** consists of a pair (P, R_s) of the set of persons and the set R_s of relations held at the date of the session s, defined as follows,

$$R_s = \{r \in R \mid date(bgn(r)) \le date(s) \text{ and } date(fini(r)) > date(s)\},$$

where $\le$ and $>$ are used to mean the order of before and after of dates. Please note that $bgn(r)$ and $fini(r)$ are the session and then $date(bgn(r))$ and $date(fini(r))$ are the date when the relation r starts and the one when ends, respectively. $network(s)$ denotes the network for a session s. The network can be a graph $G = (P, E)$, where E is given by $E = \{(from(r), to(r)) | r \in R_s\} \subseteq P \times P$, although this representation loses the other information. We call the graph $G = (P, E)$ for a network the **underlying graph** of it.

3.3 A Case as a Sequence of Scene Networks

We index sessions in S as $S = (s_1, s_2, \ldots, s_n)$ to satisfy $date(s_i) \le date(s_j)$ if $i \le j$. Then, we can have a sequence of networks $network(s_1), network(s_2), \ldots, network(s_1)$.

From the sessions $S = (s_1, s_2, \ldots, s_n)$, we take a part of the sequence $S' = (s_{i_1}, s_{i_2}, \ldots, s_{i_k})$ for $s_{i_j} \in S$ $(1 \le j \le k)$ to satisfy the following conditions.

- $i_1 = 1 < i_2 < \cdots < i_k$.
- For $j \in \{2, \ldots, k\}$, $network(s_{i'}) = network(s_{i_j})$ for any i' s.t. $i_{j-1} < i' < i_j$.

As a result we have a sequence of networks, $N = (network(s_{i_1}), network(s_{i_2}), \ldots, network(s_{i_k}))$, in which networks are different scenes in the order of time. We call a network in the sequence S' a **scene** of the case and the partial sequence N' of networks the **scene sequence** of the case.

We take this form of scene sequence as a representation of a case. In the rest of this paper we use this form and give procedures and methods to analyze cases.

4 Representation of Relations

4.1 Objective Relations and Subjective Relations

We deal with any relations about two persons, which includes, for example, the following examples, father-son, supervisor-student relation, employer-employee relation, some emotional relations, such as, respect and fears. An example of a situation is expressed by a graph in Fig. 2. Although a situation specified by relations does not represent all situations of cases (i.e., all information recorded during counseling) which the student meets, the situation can be used as an index of the difficulty.

We classify relations into two classes, objective relations and subjective relations. An objective relation is a relationship between two persons, the relationship which holds by some conditions on physical, biological or social situation but not by their mind. A subjective relationship is a relationship between two persons, the relationship holds in their mind.

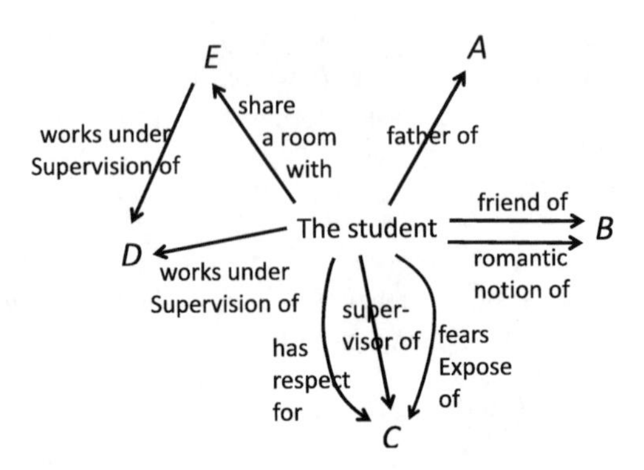

Fig. 2 Relations surrounding a student

We may be aware of that when a case includes a pair of persons including the consulter student as a relation, it includes at least a subjective relation and at least an objective relation. There is an objective relation because a person who appears in a case has to take a position or a role, such as, his/her supervisor, a friend and a parent, to the consulter student. There is a subjective relation as well because the student has some emotion to the person or has the person had some emotion. Otherwise the person need not appear in the story.

Objective relationships are very wide. They include blood relationships, relationships come from organizational roles (e.g., student-supervisor relationships, class mates, lecture teachers), relationships come from social or behavioral situations (e.g., neighbors, the relation with a person to be on the same train, relationship with a person fighting with). Subjective relationships include such as respective consideration, feeling of disgust and feeling of fear.

Classifications for objective relations can be sorted as an ontology, that is a conceptualization tool for software engineering. Katayama et al.[9] gives an ontology based on the chance which connects two persons. Fig. 3 is a part of ontology on objective relations. For subjective relations we discuss in the following section.

4.2 Attributes of Subjective Relations

Subjective relations are a class of relationships that basically exist only in mind of the person. During counseling sessions persons show emotion and feeling to other people and we need to describe subjective relations. The study of emotion has long history but it still has little consensus[1]. However, in counseling tasks we describe subjective relationships and emotional situation. As an experimental work, we give a tentative framework for subjective relationships in this paragraph.

Against the nature of the subjective relationships we need to treat them in an objective manner. From an intuitive observation we can see that subjective relations have some characters. For example a relation from a student to his mother who is very serious and always scaled her may discourage her and the relation may have negative contribution to solve the difficulty. The student herself does not desire to

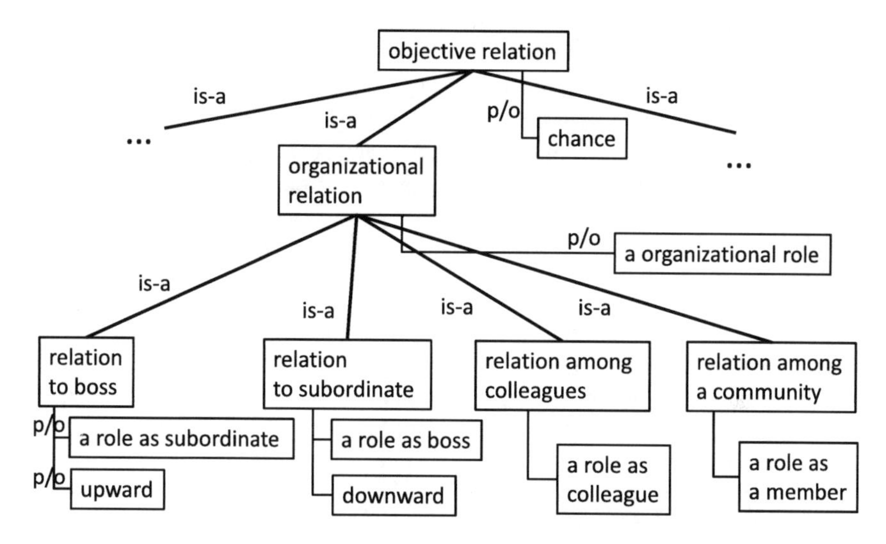

Fig. 3 A part of ontology for objective relations based on chances that connects two persons[9]

keep this relationship. These contribution and desire may help to understand the situation. We take the following five attributes including the contribution and desire to treat and represent relations. We do not intend to understand relations but give indexes for classifying. We gave the following attributes for subjective relationships.

Mode of the Emotion Defining Relations

A subjective relation accompanies an emotional feeling. Obeying the analysis by Robert Plutchik[1] we take a set of basic emotions for this attributes. Our tentative version uses *joy, love, trust, submission, fear, awe, surprise, disapproval, sadness, remorse, disgust, contempt, anger, aggressiveness, anticipation* and *optimism.*

Contribution of Relations to Solve Difficulties

We gave this attribute from the judge of a counselor. The contribution of a relation is *positive* if the relation has a positive effect in the sense that keeping the relationship contributes to solve difficulties of the student. The contribution is *negative* if the relation has a negative contribution to solve the difficulties. The value can take the value *neutral.*

Desire of the Student to Keep Relations

We gave this attribute from the position of the consulter student. The desire of a relation is *positive* if the relation is wanted to keep by the consulter student. It is *negative* if the student wants to break off the relation. Otherwise it is *neutral.*

Distance of Relations

The definition of this value is not very objective. The value for relationship from *A* to *B* is *near* if *A* feels *B* close not in geographical sense but in mental sense. If the value is *near* the student does not have barriers to contact with the alter. For the case of opposite situation the value takes *far*. It allows the value *neutral*.

Dominance of Relations

The dominance value specifies who has dominance in the relationship. The values *ego* for relationship from *A* to *B* means that *A* has dominance. The values *alter* means *B* has dominance. We give more precise definition. When with the relationship from *A* to *B* the person *A* hopes to be in the position of *B*, we say that *B* has dominance and the value is *alter*. When *A* hates to be in the position of *B*, we say that *A* has dominance and the value is *ego*.

For example let us imagine that a student has a respective emotion to a great professor whom the student takes a lecture class of but can not talk with. These relationships between the student and the professor can be characterized by the attributes. The mode of emotion is *awe*, the sign may be *negative* because the emotion is too strong, the direction is *yes*, the distance is *far*, and the dominance is *alter*.

For another case let us imagine a relationship from a girl to her friend girl. The both are always together and have strong dependence each other. The mode of emotion for this relationship is *trust*, the sign may be *negative*, the direction is *no* because it is symmetric, the distance is *near*, and the dominance is *neutral* or possibly *alter*.

5 Analyzing Cases

This section presents cases, which are anonymous and changed from real cases but include real structure of counseling. We need careful treatment for studying counseling cases because they are extremely confidential. We have treated cases by the following steps.

1. A counselor takes records for counseling sessions.
2. After closing sessions the counselor takes consent with consulter students to use their records as research material when it is possible.
3. The records are carefully made anonymous by removing all information that connect to identify students.
4. The anonymous records are changed and rearranged into fictional cases which are not real but include real elements and structure.

The steps 1 and 2 are only done by a professional counselor. Step 3 is done by an automated system which is a function of the recording system that our college developed and by hand. Step 4 is done by hand of one of authors who is in charge of student advising section under the supervision of the professional counselor. The other authors of this paper, who are not professional in counseling matter, use the records after all steps. Some cases are taken from published books for counseling studies.

We present four cases in Table 1. These are typical medium size records. Table. 2 is a summary of sizes of cases, the number of sessions, the number of scenes, the number of persons and the number of relations.

Table 1 Outlines of example counseling cases

Case a : A male undergraduate student isolated from college for two years came the counseling desk with his mother, the mother who did not know the situation and were confusing. At the attendance the student was not willing to be back the college. By counseling the student became to talk his interest and became have will to study again. He gained relationships with friends. Finally he finished graduation and found a job.

Case b : A male undergraduate student attended with a complaint about his health and mental problems. Cooperation with a clinic outside of college he continued to have counseling. During the period a counselor gave him a chance of work therapy and brought him to involve many experience. Accordingly he gained his confidence and successfully he joined research project in a laboratory.

Case c : A male graduate student distressed himself because continuing fail of job position and it brought a slump in his research project. He stopped to attend the laboratory by the stress from his supervisor professor. The counselor advised him to see himself positive acceptance and asked head of department to rearrange the supervision for him. Temporally his depression come his disappearance but he recovered his motivation with warm treatment of his family. Then he finished his course.

Case d : A female graduate student had a complaint with a serious expression about sexual contact by a staff of her laboratory. She made complaint to the staff but it caused a worse situation. Her distress brought symptoms in her health. Students of the same laboratory did not help her. The counselor advised to attend a clinic and continued counseling. After recovering her health the counselor gave her a chance to meet other students who can help her. Although she stayed in her laboratory, she managed to avoid the annoying situation. Finally she finished her study successfully.

Table 2 Basic information of the cases presented in Table 1

case	#sessions	#scenes	#persons	#relations
case a	40	11	6	32
case b	31	10	5	28
case c	16	6	5	37
case d	12	10	9	37

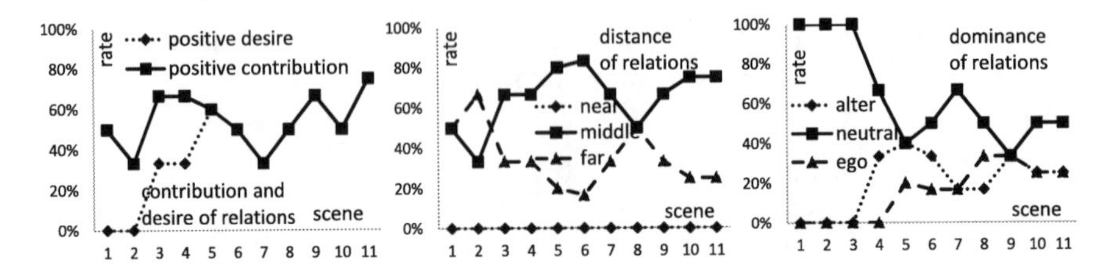

(a) Transitions of relation attributes in case a

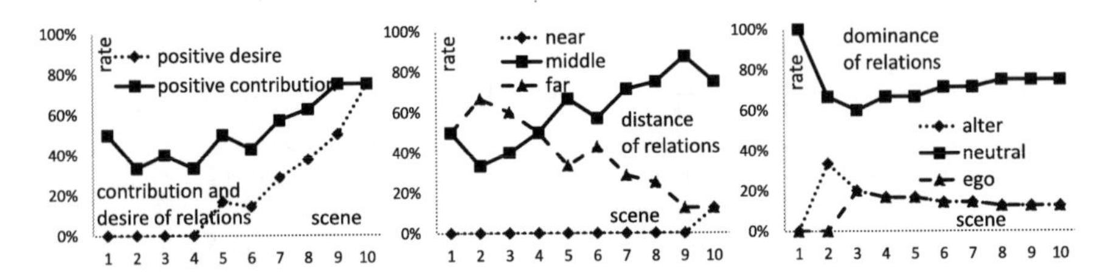

(b) Transitions of relation attributes in case b

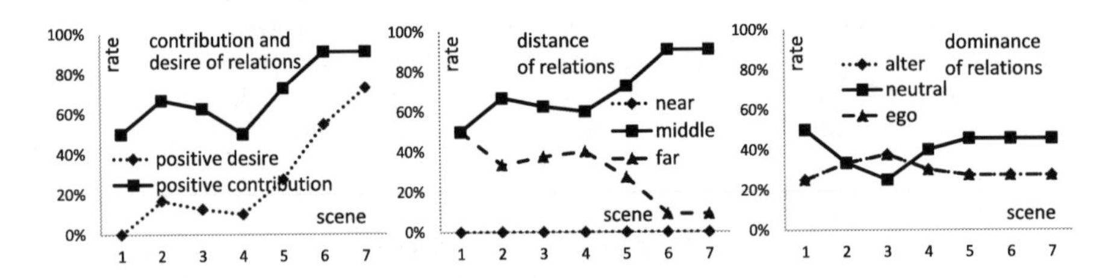

(c) Transitions of relation attributes in case c

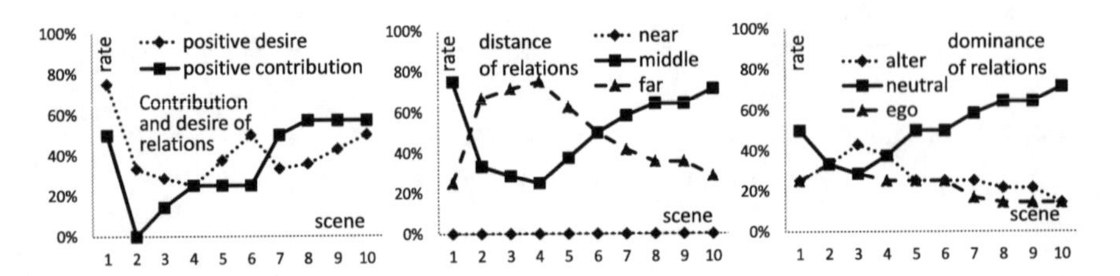

(d) Transitions of relation attributes in case d

Fig. 4 Transitions of attributes' values of cases in Table 1. For a case a line includes graphs for desire and contribution attributes, for distance and for dominance.

Four cases had more or less happy ending, although cases were not necessarily resolved. As a preliminary analysis of cases, we show transitions of attributes in Fig. 4. Four lines are for four cases. The left graphs of lines show the transitions of the rates of positive contribution and positive desire. The lines have tendency of increasing rates according as scene goes. The center graphs show the rates of near, medium, and far distance of relations. They show the tendency of decrease of far distance, which means development of closer and dense networks. The right graphs are the transitions of the rates of dominance values. If the value takes *ego* it means an possibility of selfish behavior. The tendency is not clear but the transitions may show an evolution of mutual understanding.

6 Processing Cases

In order to support the practical operations of student counseling, this section tries to give some computational methods. Considering routine of use of a counseling record system, searching function of records is essential. For the advanced application clustering of cases may help to understand situations of students in difficulty. Accordingly we will define similarity measure among cases for these purposes.

For given two cases c_1 and c_2 we will define the similarity measure $sim_{\text{case}}(c_1, c_2)$. We may understand two cases are similar when two stories of cases have similar lines, and persons and their roles and emotional relations correspond, each other. Hence we try to define similarity by the degree how far two cases are matched in networks and story lines. In the rest of this section we give definitions of similarities between two relations, between two persons, between two scenes, and between two cases.

6.1 Similarity between Two Relations

Among relations we give definitions of similarity for between two objective relations and between two subjective relations separately. Since we gave an ontology for objective relations, we use a similarity measure defined for ontology diagrams. Based on the positions of concepts for relations in the ontology diagram, similarity measures are proposed, such as, in [7]. We used an ontology editor Hozo[5] to develop the conceptual ontology for objective relations. The editor Hozo also gives an similarity measure based on the topology of ontology diagrams. We obey and applied the similarity measure of Hozo for the similarity measure among objective relations. We denote the similarity between two objective relations r_1 and r_2 by $sim_{\text{ob-rel}}(r_1, r_2)$.

For two subjective relations r_1 and r_2 we give similarity measure $sim_{\text{sb-rel}}(r_1, r_2)$ by the accordance of attributes as follows,

$$sim_{\text{sb-rel}}(r_1, r_2) = \frac{\sum_{\text{attribute } a}(1 - dist(a(r_1), a(r_2)))}{\text{the number of attributes of subjective relations}}, \quad (1)$$

where $a(r)$ means the value of the attribute a of the relation r and $dist(a(r_1), a(r_2))$ is 1 if the values of an attribute a for r_1 and r_2 disagree, 0 if agree, 0.5 if one of values is *neutral*. We use the attributed, contribution, desire, distance and dominance, defined in the previous section for the similarity of subjective relations. The similarity does not count for the mode of relation.

As a result we define the similarity $sim_{\text{rel}}(r_1, r_2)$ of two relations as follows.

$$sim_{\text{rel}}(r_1, r_2) = \begin{cases} sim_{\text{ob-rel}}(r_1, r_2) & \text{if both } r_1 \text{ and } r_2 \text{ are objective} \\ sim_{\text{sb-rel}}(r_1, r_2) & \text{if both } r_1 \text{ and } r_2 \text{ are subjective} \\ 0 & \text{otherwise} \end{cases} \quad (2)$$

6.2 Similarity between Two Person Pairs

When we choose a pair of persons from a scene and another pair from another scene we need to measure the similarity. A difficult thing to define the similarity for pairs is that there are more than one relation among persons and the numbers of relations may different.

We consider a pair (p_1, p_2) of persons from a case and another pair (q_1, q_2) from another case. Then let $Rel(p, p')$ denote the set of all relations between p and p'. We assume that $|Rel(p_1, p_2)| \leq |Rel(q_1, q_2)|$ and imagine an injection, i.e. one-to-one mapping, from $Rel(p_1, p_2)$ to $Rel(q_1, q_2)$, which give a correspondence between relations held for two pairs.

Thus, the similarity $sim_{\text{person}}((p_1, p_2), (q_1, q_2))$ between pairs are given by,

$$sim_{\text{person}}((p_1, p_2), (q_1, q_2)) = \begin{cases} \max_{f \in F} \operatorname{average}_{r \in Rel(p_1, p_2)} sim_{\text{rel}}(r, f(r)) \\ \quad \text{if both pairs have relations} \\ 1 \quad \text{if neither have relations} \\ 0 \quad \text{if one has relations and the other does not,} \end{cases} \quad (3)$$

where F is the set of all injection from $Rel(p_1, p_2)$ to $Rel(q_1, q_2)$.

6.3 Similarity between Two Scenes

In order to correspond a scene to another, we need a matching of persons in scenes. We have to be careful for that the correspondence of a person to a person has to be fixed through the scenes in matching of two cases. Accordingly we imagine the accumulated networks through a case. Let $N = (network(s_{i_1}), network(s_{i_2}), \ldots, network(s_{i_k}))$, where $network(s_{i_j}) = (P, R_{s_{i_j}})$, be a scene sequence of the case. Then an **accumulated network** of the case is the pair (P, R) where $R = \cup_{j \in \{1,\ldots,k\}} R_{s_{i_j}}$.

Thus, when we have two accumulated networks (P, R) and (Q, T) for two cases. We assume $|P| \leq |Q|$ without lost of generality. We may have a correspondence of persons by an injection from P to Q that maximize the correspondence between R

and T. That is, we may give a definition of the similarity between two scenes N_1 and N_2 by,

$$sim^{\hat{g}}_{\text{scene}}(N_1,N_2) = \operatorname*{average}_{(p_1,p_2)\in E} sim_{\text{rel}}((p_1,p_2),(\hat{g}(p_1),\hat{g}(p_2))), \quad (4)$$

where $\hat{g} = \textit{max-injection}(N_1,N_2)$ which is defined by

$$\textit{max-injection}(N_1,N_2) = \operatorname*{argmax}_{g\in G} \operatorname*{average}_{(p_1,p_2)\in E} sim_{\text{rel}}((p_1,p_2),(g(p_1),g(p_2))) \quad (5)$$

and $E \subseteq P \times P$ is the edge set of the underlying graph of the accumulated network (P,R), and G is the set of all injections from P to Q satisfying that the consulter student in P has to be mapped into the consulter student of the other case in Q.

6.4 Similarity between Two Cases

Finally we will define the similarity of two cases, or two scene sequences. Let $N = (n_1,\ldots,n_k)$ and $M = (m_1,\ldots,m_{k'})$ be two scene sequenced for two cases c and d and we assume $k \leq k'$. Then the similarity between the two cases is defined as follows,

$$sim_{\text{case}}(c,d) = \max_{h\in H} \operatorname*{average}_{i\in\{1,\ldots,k\}} sim^{\hat{g}}_{\text{scene}}(n_i,m_{h(i)}), \quad (6)$$

where $\hat{g} = \textit{max-injection}(N_1,N_2)$ and H is the set of functions from $\{1,\ldots,k\}$ to $\{1,\ldots,k'\}$ satisfying that $h(i) < h(j)$ if $i < j$.

6.5 Approximation of Similarity Computation and Application

The definition of similarity of two cases includes many matching and maximization steps. In order to have practical procedure we introduce an efficient technique and some approximation procedures for the similarity computation. We used a greedy technique to match relations, to choose the best injection f for Equation (2), and also the best injection g for Equation (3).

The greedy procedure for f, which is an injection function from the set $Rel(p,p')$ of relations held between a pair (p,p') of persons in a case to $Rel(q,q')$ between another pair in another case, takes a maximally similar pair of relations $(r,r') \in Rel(p,p') \times Rel(q,q')$ for the first matching. Next, it takes the maximally similar pair from $(Rel(p,p') - \{r\}) \times (Rel(q,q') - \{r'\})$. It continues until every $r \in Rel(p,p')$ is matched to a relation in $Rel(q,q')$.

The greedy procedure for g works likewise. Let P and Q ($|P| \leq |Q|$) be the sets of persons of two cases. The procedure matches the consulter students in P and Q at first and keeps these two persons f_P and f_Q in focus. Then, it considers all persons in P who have relations with f_P, that is $P' = \{p \in P \mid f_P \text{ has a relation to } p\}$. It also considers $Q' = \{q \in Q \mid f_Q \text{ has a relation to } q\}$. Then the procedure takes the maximum similar two pairs from (f_P,p) s.t. $p \in P'$ and (f_Q,q) s.t. $q \in Q'$. The chosen maximally similar two persons from P' and Q' become the next focused persons f_P and f_Q. It continues to choose maximally similar two persons who are related by f_P and f_Q, respectively. If it fails the procedure takes back track to the previous focused persons. It continues until every person in P has correspondence to a person in Q.

For the matching of the sequence, that is, choice of h in Equation (6) we do not need approximation. We can use the efficient matching algorithm of DP-matching, which was originally developed for speech recognition and has been used in many fields.

We implemented the procedure of similarity computation. In order to evaluate efficiency of the procedure we need to prepared many cases. As we mentioned, cases are strictly confidential and it is difficult to prepare many cases. To overcome this difficulty we developed a procedure to manipulate and produce fictional cases from a small number of given cases. The manipulations include inserting a random relation, deletion of relation, changes of start or ending relations, changes of attribute values. These manipulations can be regarded as a noise adding. If we takes the sum of total number of attributes of relations in scenes in a case as its volume, we can measure the rate of noise by the rate of the noise added in the volume.

When we prepared 100 cases and tried to find ten the most similar cases to a case. This operation took 2.3 seconds, which is a practical run time. When we computer the similarity by complete matching without the approximation the computation took 8.5 seconds. The result of the most similar cases by the complete computation and approximation were the same.

7 Concluding Remarks

In this paper we discussed the possibility and necessity of contribution of IT and AI technology to student counseling activities. We challenged this theme by analyzing the structure of counseling records and gave the database schema and mathematical formal representations. The representation is based on the idea that counseling cases are regarded as a sequence of networks among persons, of which each different network is called a scene. According to these formalization we tried to give a systematic classification of relations by an ontology and attributes. We demonstrated that these ideas has potential to be used to understand counseling cases and also to process them, i.e. searching and clustering, for practical routines.

Our future works include the development an effective interface to edit the information, i.e. relations and their attributes. Our counselor suggested that the documentation including our proposal does not disturb the busy work of the counselor, because these information is already documented in text and an effective interface may accelerates the documentation routine. The attributes of the subjective relations are given to be objective. Our counselor evaluates the attributes are adequate and she suggests other improvement. We need to examine large number of cases with the attributes.

References

1. Plutchik, R.: The Nature of Emotions. American Scientist 89, 344–350 (2001)
2. Trow, M.: Reflections on the Transition from Mass to Universal Higher Education. American Academy of Arts and Sciences (1970)
3. Ramakrishnan, N., Hanauer, D., Keller, B.: Mining Electronic Health Records. Computer 43(10), 77–81 (2010)
4. Schuh, J.H., Jones, S.R., Harpe, S.R., associates: Student Services — A Handbook for Profession, 5th edn. Wiley (2003)
5. Mizoguchi, R., Sunagawa, E., Kozaki, K., Kitamura, Y.: A Model of Roles within an Ontology Development Tool: Hozo. Applied Ontology 2(2), 159–179 (2007)
6. Oberle, D., Grimm, S., Staab, S.: An Ontology for Software. In: Handbook on Ontologies, 2nd edn., pp. 383–402. Springer (2009)
7. Ehrig, M., Haase, P., Hefke, M., Stojanovic, N.: Similarity for Ontologies - a Comprehensive Framework. In: Workshop Enterprise Modeling and Ontology: Ingredients for Interoperability, at PAKM 2004 (2004)
8. Sakoe, H., Chiba, S.: Dynamic programming algorithm optimization for spoken word recognition. IEEE Transactions on Acoustics, Speech and Signal Processing 26(1), 43–49 (1978)
9. Katayama, T., Oda, N., Mutoh, A., Inuzuka, N.: Ontology of Human Relationships - An Approach to Computer-Aided Student Counseling. In: Graña, M., et al. (eds.) Advances in Knowledge-Based and Intelligent Information and Engineering Systems - 16th Annual KES Conference, Frontiers in Artificial Intelligence and Applications, vol. 243, pp. 1788–1796. IOS Press (2012)

Web Based UNL Graph Editor

Khan Md Anwarus Salam, Hiroshi Uchida, Setsuo Yamada, and Nishio Tetsuro

Abstract. The Universal Networking Language (UNL) is an artificial language which enables computers to process knowledge and information across language barriers and cultural differences. Universal Words (UWs) constitute the vocabulary of UNL. However, from the UNL expression it is difficult for human to visualize the UNL graph to edit it interactively. To solve this problem, we have developed the web based UNL graph editor which visualizes UNL expression and the users can edit the graphs interactively.

Keywords: UNL, Visualization, NLP, Web.

1 Introduction

The Universal Networking Language (UNL) is an artificial language which enables computers to process knowledge and information across language barriers and cultural differences. Using UNL computers enable humans to communicate with each other and share information and knowledge using their native language. UNL initiative was originally launched in 1996 as a project of the Institute of Advanced Studies of the United Nations University (UNU/IAS). Describing the detail technical information UNL book was first published in 1999 [1]. In 2001, the United Nation University set up the UNDL Foundation, to be responsible for the development and management of the UNL project. In 2005, a new technical manual of UNL was published [2], which defined UNL as a knowledge and information representation language for computer.

Khan Md Anwarus Salam · Nishio Tetsuro
Graduate School of Informatics and Engineering,
The University of Electro-Communications, Tokyo, Japan
e-mail: kmanwar@gmail.com, nishino@uec.ac.jp

Hiroshi Uchida
UNDL Foundation, Tokyo, Japan
e-mail: uchida@undl.org

Setsuo Yamada
NTT Corporation, Tokyo, Japan
e-mail: yamada.setsuo@lab.ntt.co.jp

R. Lee (Ed.): *SNPD*, SCI 492, pp. 219–228.
DOI: 10.1007/978-3-319-00738-0_16

However, it is difficult for human to visualize the UNL expression as a graph. Previously there was a desktop based UNL visualization tools available. But it could not visualize the UNL scope as one node. There is a web based tool for visualizing the UNL Ontology [5]. But it could not express the UNL graph as the visualization techniques are different from UNL Ontology and UNL graph.

It would be helpful for the UNL experts to have a web based UNL graph editor with a graphical interface. To solve this problem, we have developed the web based UNL graph editor which visualizes UNL expression and the users can edit the graphs interactively.

The main features of our web based UNL editor are UNL visualization, UNL graph in different languages, better way to represent UNL scope and shows UWs explanation in different languages.

This UNL graph editor can be used as the core tool for visualizing UNL expression into any web based application for UNL technologies. The users of the system are researchers and linguists who are interested to work with UNL. The purpose of the system is to help the users to provide visual representation of the UNL graph for better understanding. The system solves the problem of the users to go through the raw data for getting information from the UNL texts. Users can use this web system using popular web browser with internet access. In this paper we described the web based system for visualizing the UNL graphs. This UNL graph viewer can be accessed using web browsers.

2 Previous Way of Editing

UNL, as a language for expressing information and knowledge described in natural languages, has all the components corresponding to that of a natural language. Universal Words (UWs) constitute the vocabulary of UNL. Each concept that natural languages have can be represented as unique UW. A UW of UNL is defined in the following format:

<uw> =:: <headword>[<constraint list>]

Here, English words or phrases are used for headword, because of easy understanding for the people in the world. UW can be a word, a compound word, a phrase or a sentence. Universal Words (UWs) constitute the vocabulary of UNL. UW is a word for constructing UNL expressions (UNL Graph). So keys to the information in UNL documents are UWs. UWs are stored in the UW dictionary. UWs are inter-linked with other UWs using “relations” to form the UNL expressions of sentences. These relations specify the role of each word in a sentence. Using "attributes" it can express the subjectivity of author. Each UWs are interlinked with each other through the UW System in the UNL Ontology. Master definitions for UWs describe all relations that a UW can hold. A minimum set of relations is used as constraints of UW for the purpose to make a UW distinguishable from sibling UWs.

UNL Ontology is a lattice structure where UWs are inter-connected through relations including hierarchical relations such as icl (a-kind-of) and iof (an-instance-of). UNL Ontology includes possible relations between UWs, UWs definition and UNL system hierarchy. In the UNL Ontology, all possible semantic co-occurrence relations, such as 'agt', 'obj', etc, between UWs are defined based on the UW System. Every possible semantic co-occurrence relation is defined between the two most general UWs in the hierarchy of the UW System that can have the relation. With the property inheritance characteristic of the UW System, possible relations between lower UWs are deductively inferred from their upper UWs and this inference mechanism reduces the number of binary relation descriptions of the UNL Ontology. In the topmost level UWs are divided into 4 categories adverbial concept, attributive concept, nominal concept and predicative concept.

UNL expresses information or knowledge in the form of semantic network with hyper-node. In the UNL semantic network, nodes represent concepts and arcs represent relations between concepts. Concepts can be annotated. Such semantic network is called "UNL expression".

UNL expression is a semantic network made up of a set of binary relations, each binary relation is composed of a relation and two UWs that hold the relation. A binary relation of UNL is expressed in the following way:

<relation> (<uw1>, <uw2>)

In <relation>, one of the relations defined in the UNL Specifications is de-scribed. In <uw1> and <uw2>, the two UWs that have the relation given by <relation> are described. A semantic network of UNL expression is a directed graph composed of binary relations with direction. The three elements of a binary relation have the following interrelationship: <uw1> -- <relation> -> <uw2>

This binary relation is interpreted as that: the UW given in <uw2> plays the role indicated by the relation given in <relation> held by the UW given in <uw1>; whereas the UW given in <uw1> holds the relation given in <relation> with the UW given in <uw2>

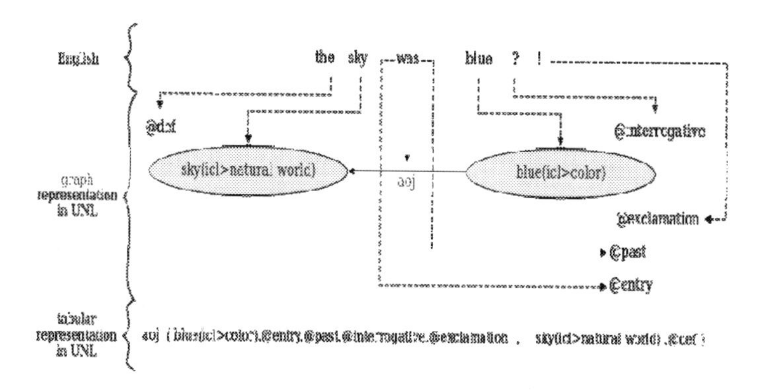

Fig. 1 UNL expression example explanation

For example, the English sentence 'The sky was blue?!' can be represented using the following UNL expression:

{unl} aoj(blue(icl>of the blue color(aoj>thing)):0C.@entry.@past .@interrogative. @exclamation, sky(icl>natural world):04.@def) {/unl}

To check UNL information users need to navigate through the relations of UWs given in UNL expression. This is difficult for the users to browse those UWs in UNL ontology. Moreover, It is troublesome for the beginners to find the UW relations in raw data format. To address these problems, we have designed an interactive web-based system for UNL graph visualization. Instead of leaving the users to manually traverse the UNL graph, the developed system lets user visualize the UNL expression into a graph.

The interactive ontology visualization encodes a number of properties that help users to see the relations of UWs to get the semantic background. And users remain oriented while navigating the ontology through the web browser. It helps the users to provide visual representation of the ontology for better understanding. The system can be accessed from any popular web browser such as Internet Explorer, Mozilla Firefox etc. The system also requires the browser to support Javascript.

The input of our system is the UNL expression and the output is the interactive visualization of the UNL graph. We developed a UNL parser which finds all the UWs and relations between then. Then the system draws the graph using SVG technology. Moreover each UWs node are connected with the UNL ontology. It enables users to immediately learn about the particular UW. For example if the user wants to know about the UW "blue(icl>of the blue color(aoj>thing))", they can directly access the UW properties from UW dictionary using the same interface.

3 Web Based UNL Editor

The main features of our web based UNL editor are UNL visualization, UNL graph in different languages, better way to represent UNL scope and shows UWs explanation in different languages.

3.1 UNL Visualization

From raw data, it is difficult for human to visualize the UNL graphs. This paper described the web system to visualize UNL graphs. It will help the users of UNL by providing an interactive web interface to work with UNL graphs. Figure2 shows a sample screenshot of the UNL graph, which visualize the UNL expression for the English sentence 'The sky was blue?!'. Here the grey circles represent the UWs with the label of the UWs headword in grey color and UW id in red color. The reason behind not to show the constraint list is to simplify the graph view even for the beginners of UNL.

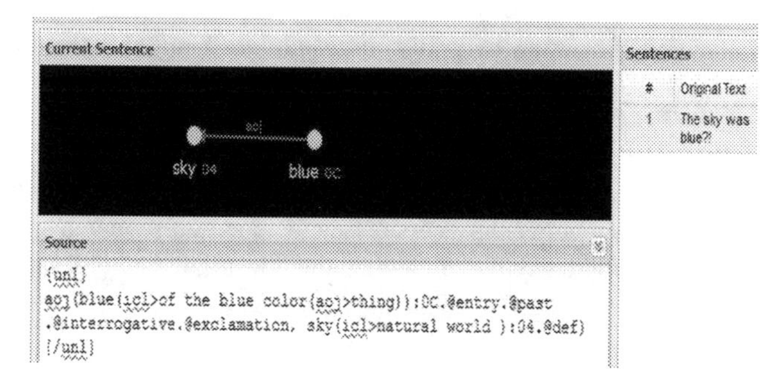

Fig. 2 Screenshot of the web based UNL graph editor

However, the constraint list of UW is also important for UNL experts. If require in our system users can see the full UW with the constraint list and attributes in a tool tip box. Figure3 shows such an example where it shows the full UW with the constraint list and attributes for the node "blue".

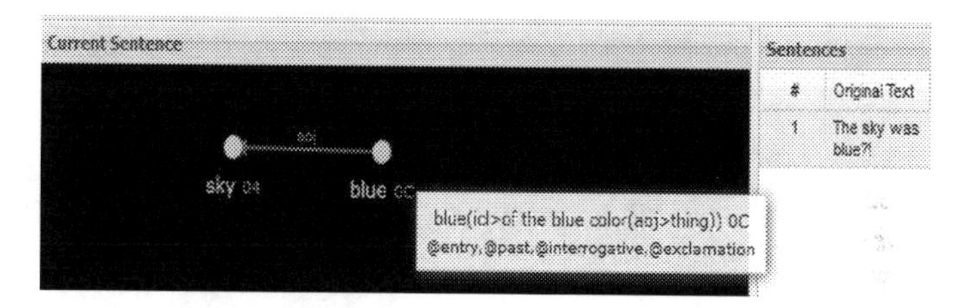

Fig. 3 Screenshot of the UW tooltip

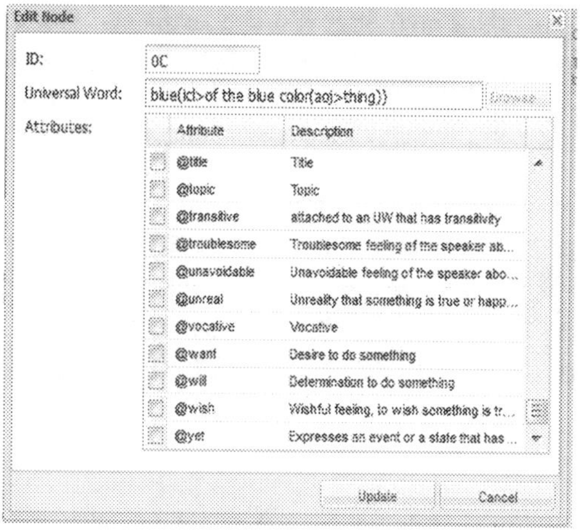

Fig. 4 Screenshot of the Edit node window

Users can also edit the nodes, by performing right click in the node and then by selecting “Edit Node” option. Then the Edit node window will be opened as shown in Figure4. Here users can edit the node ID, UW or its attributes with simple clicks.

UNL has around 46 relations. In our graph editor we used different color code for different relations. It helps the users to easily distinguish between different relations. Current system shows one UNL sentence in a graph. For example, Figure 5 shows one sentence with multiple color coded relations.

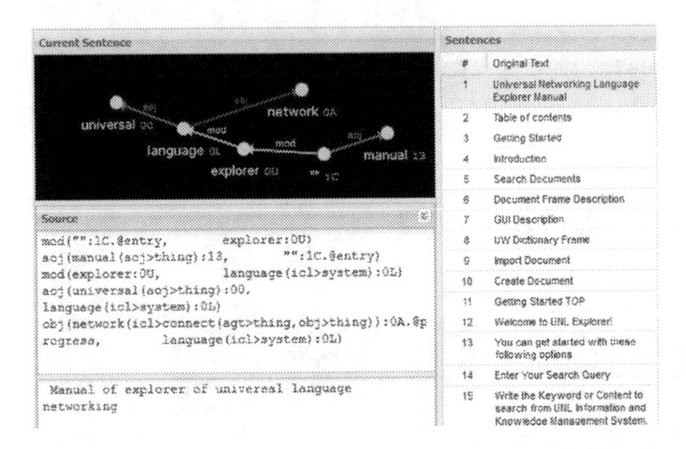

Fig. 5 Color coded relations

3.2 *UNL Graph in Different Languages*

The unique feature of this graph editor is that using the UWs dictionary it can display one UNL graph into different languages. This is the unique feature of this system. The figure 6 shows the example of UNL graph displaying in Bengali language. It shows the usefulness of such system for the native speakers of that language.

Fig. 6 UNL graph of “blue sky” in Bengali

The figure 7 shows the example of UNL graph displaying in Japanese language. The native speakers of Japanese language can be benefitted by this system when they need to understand the UNL graph.

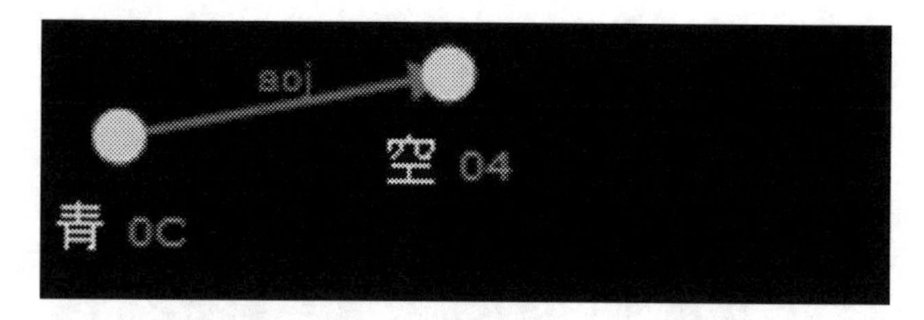

Fig. 7 UNL graph of "blue sky" in Japanese

3.3 Better Way to Represent UNL Scope

In UNL expression, a "scope" is a group of relations between nodes that work as a single semantic entity. For example, in the sentence "Khan saw Rupok when Ratan arrived", the dependent clause "when Ratan arrived" describes the argument of a time relation and, therefore, should be represented as a hyper-node or as a sub-graph. This kind of hyper graph structure is essential to express natural languages using graph. However, it is very important to have a meaningful way in graph editor to visualize the scope nodes. In general scope node normally express a concept that expresses a phrase or a clause in natural language. We need to provide the way to express of this compound concepts in the UNL graph.

In our web based UNL editor we can display the UNL scopes as hyper nodes. Hyper nodes has very strong structure to deal with natural language. Previous graph editors could not visualize the scopes properly. As UWs are kind of symbols for computers to distinguish among concepts, we need to show the concepts using nodes. It is required to visualize the UWs expressed using UNL syntax. For example figure 8 shows the simple view of UNL graph. However, each 01 or 02 nodes contains another sub graph or the UNL expression of natural languages.

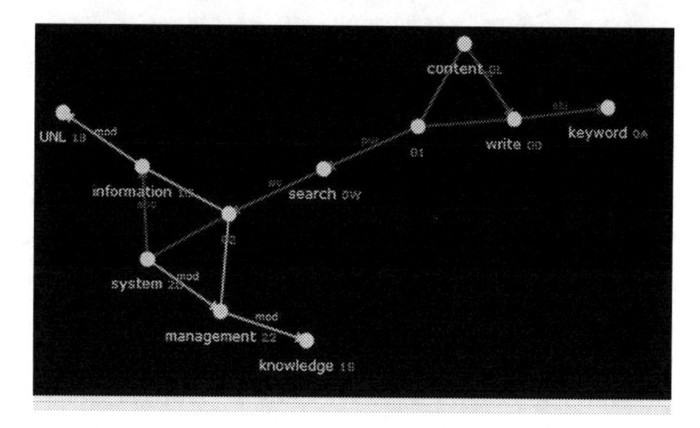

Fig. 8 UNL scopes in simple view

To solve the problem of visualizing the scope, we propose to translate the scope UNL expression. For example, figure8 is the most simplified expression of UNL graph in recent years. Figure9 shows the scopes translation part done by computer.

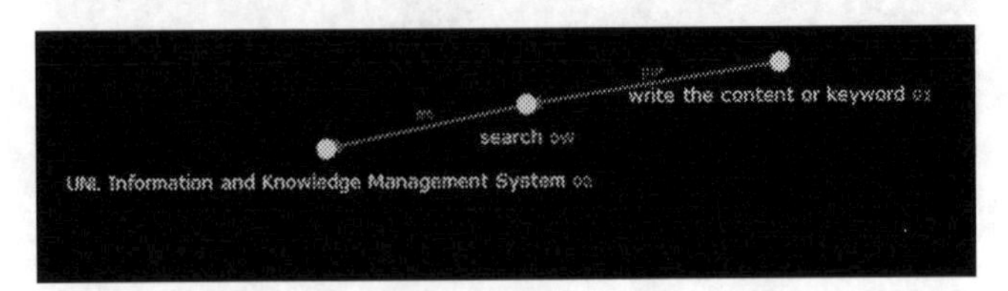

Fig. 9 Scope graph in Japanese language

Using UWs dictionary it is possible to show the UNL graph in any language as well. All these graphs are helpful for understanding the UNL representation of the language. Using the same mechanism we an also translate the graph in to Figure10.

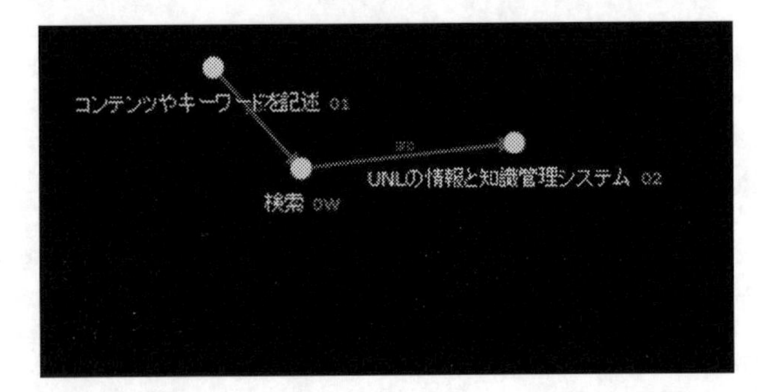

Fig. 10 Scope graph in Japanese language

Moreover, UWs Master definition editors can express the UWs definition. For example, to visualize Scope 01 and 02, both scopes can be expressed in English or any other language using the UNL Deco for that UNL expression of 01 and 02.

3.4 Shows UWs Explanation in Different Languages

This graph editor can partly show the UWs explanation in different available languages which has been first proposed before to help the UNL dictionary builder [3]. Moreover, UWs itself are formal and not always to be understandable for human. To ensure every language speakers can create the correct UWs dictionary entry, we need to provide the explanation of UWs in different natural languages for humans. As there are millions of UWs, it is very expensive to manually build the UWs explanation in all natural languages. To solve this problem, this research proposes the way to auto generate the UWs explanation in UNL, using the property inheritance based on UW System. Using UNL DeConverter from that UNL the system can generate the explanation in more than 40 languages.

Figure 11 shows the UWs explanation generation steps. The input of this system is one UW and the output of the system is the explanation of that UW in natural language such as in English. For the given UW, the system first discover a SemanticWordMap [4], which contains all direct and deductively inferred relations for one particular UW from the UNL Ontology. So input of this step is one UW and output of this step is the WordMap graph. In next step we convert the WordMap graph into UNL using conversion rules. This conversion rules can generate "From UWs only" and "From UNL Ontology", based on user's requirement. So input of this step is the WordMap graph and Output is the UNL expression. In the final step we describe in natural language by converting the UNL expression using UNL DeConverter, provided by UNL Explorer.

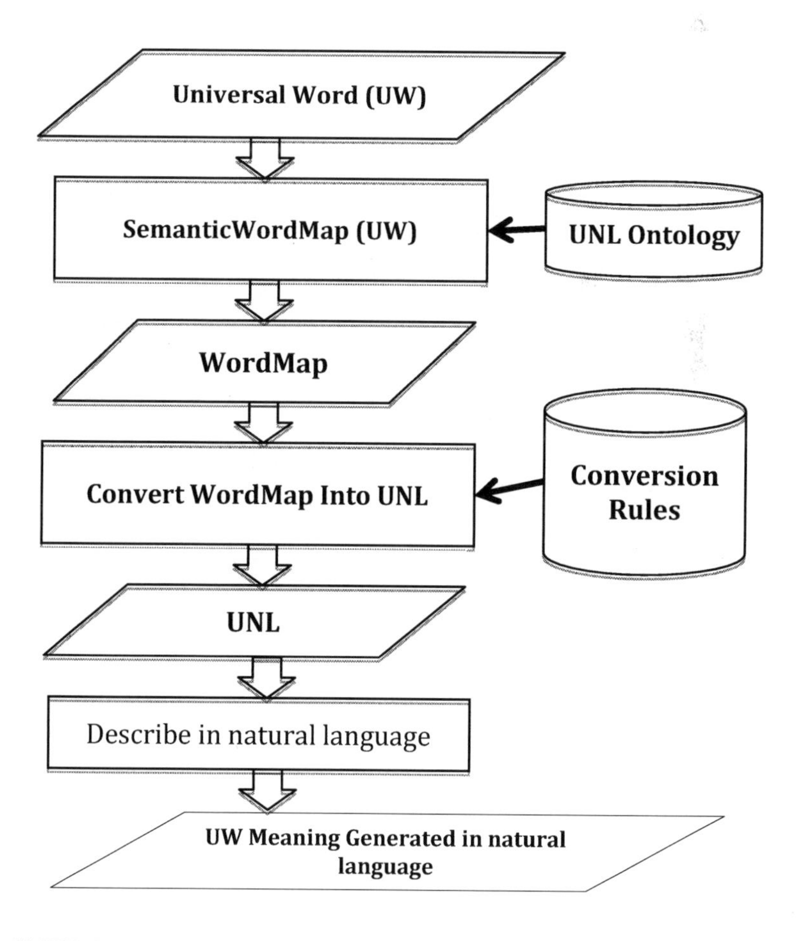

Fig. 11 UWs Explanation Generation

4 Conclusions

This paper described the web system to interactively edit the visualized UNL graphs. Before from raw UNL expression, it was difficult for human to visualize the UNL graphs online. This system will help the users of UNL system by providing an interactive web interface to work with UNL graphs. Beside the UNL graph visualization it also provides the mechanism to express the UNL graph in different languages and generate the explanation for the UWs in natural languages. It helps to build the bridge between the UNL system and human, for building the UNL resources. In future this UNL graph editor will provide API for other web based applications as well. This API will enable any web application to visualize UNL expressions with interactive editing interface. Using all these features the world language communities can use and utilize the UNL resources.

References

1. Uchida, H., Zhu, M., Della Senta, T.: A gift for a millenium. IAS/UNU, Tokyo (1999)
2. Uchida, H., Zhu, M., Della Senta, T.: The Universal Networking Language, 2nd edn. UNDL Foundation (2005)
3. Anwarus Salam, K., Uchida, H., Nishino, T.: Multilingual Universal Word Explanation Generation from UNL Ontology. Cognitive Aspects of the Lexicon, CogaLex (2012)
4. Khan, A.S., Uchida, H., Nishino, T.: How to develop universal vo-cabularies using automatic generation of the meaning of each word. In: NLPKE 2011, pp. 243–246 (2011)
5. Khan, A.S., Uchida, H., Yamada, S., Nishino, T.: UNL Ontology Visualization for Web. In: SNPD 2012, pp. 542–545 (2012)

Tracking Multiple Objects and Action Likelihoods

Chi-Min Oh and Chil-Woo Lee

Abstract. In this paper we propose a method which can improve MRF-Particle filters used to solve the hijacking problem; independent particle filters for tracking each object can be kidnapped by a neighboring target which has higher likelihood than that of real target. In the method the motion model built by Markov random field (MRF) has been usefully applied for avoiding hijacking by lowering the weight of particles which are very close to any neighboring target. The MRF unary and pairwise potential functions of neighboring targets are defined as the penalty function to lower the particle's weights. And potential function can be reused for defining action likelihood which can measure the motion of object group.

Keywords: Multiple Object Tracking, Action Likelihood, Gesture Recognition.

1 Introduction

Data association for tracking multiple objects has been an important issue because the maintenance of the list of the detected objects is needed in the research fields of image and signal processing, computer vision, stochastic process and pattern recognition. The representative applications of tracking multiple objects is such as visual player tracking in soccer broadcast system, visual surveillance in building or with mobile robots, automobile's collision avoidance against pedestrians and bicycles, multiple touch tracking for mobile devices, activity analysis of insects or people, video compression, and human-computer interaction.

The process of data association is like the process of multiple object tracking that assigns proper identities to the detected objects in the consecutive video

Chi-Min Oh · Chil-Woo Lee
School of Electronics and Computer Engineering,
Chonnam National University, Gwangju, Korea
e-mail: sapeyes@image.chonnam.ac.kr, leecw@chonnam.ac.kr

R. Lee (Ed.): *SNPD*, SCI 492, pp. 229–240.
DOI: 10.1007/978-3-319-00738-0_17

images. By only using object detectors such as a labeling method on object segmentation image not considering sequential consistency of object identities, the list of detected objects changes at every frame. There must be frequently and repeatedly appearing objects and it is required to assign same identities to these sequentially existing objects. Multiple hypothesis tracking [1] and the joint probabilistic data association filter (JSDAF) [2] are the basic frameworks in this data association area. The multiple hypothesis tracker and JPDAF with a particle filter [3] show some potentialities for nonlinear systems. However, the hijacking problem has not been explicitly resolved. Some related works [1,3,6,7] focused on other aspects, such as the observation model and feature descriptors.

Khan et al. [4] first have attempted to explicitly represent the hijacking problem. Khan uses a graph representation of neighboring objects since hijacking happened when the locations of the objects were very close. Considering the distance between objects, neighboring objects are connected as graph edges. This graph can be assumed as Markov random field (MRF) where the properties of neighboring objects can be defined as the potential functions and exploited for object tracking problem. Using MRF, a description of neighboring objects, Khan defined MRF motion model in Markov chain Monte Carlo-based particle filter and solved hijacking problem. Based on MRF motion model, to solve the hijacking problem, Khan used a penalty function from MRF pairwise potential functions to lower the particle weight nearly approached to neighbors than object itself.

This paper basically adapts Khan's MRF motion model in our multiple particle filters which we call here as MRF-Particle filters. The pairwise term of MRF-Particle filter for the penalty function is similar to Khan's method but we try to additionally utilize the unary term because the graphs of neighboring objects are built at every end of frame for penalty function in next frame that means the actual penalty function is not made at current time. Therefore unary term could compensate the weakness of pairwise term which is based on the outdated graph by one time step. Unary term can be used in penalty function with current observation image; we can add current information to one-frame outdated graph for hijacking problem.

MRF graphs represent the details of movements of all objects. Based on the potential functions of MRF graphs, it is possible to know the motion patterns of multiple objects. We present how to define the action likelihoods of neighboring multiple targets. The action likelihoods represent the motion and chords of touched fingertips based on MRF potential functions. The motion and chord likelihoods are such as clues for translation, rotation, scaling motions and structure of multiple touch points. These likelihoods can be estimated concurrently as a joint likelihood to define more complex gesture recognition. In applications, touch command can benefit from this definition strategy. By multiplying any action likelihood as a joint likelihood [9] it is possible to generate the gesture recognition command.

The rest of this paper is organized as follows. Section 2 outlines from the object detection to data association of multiple particle filters. Section 3 describes how to build MRF graphs of multiple objects. Based on MRF, the penalty function with pairwise and unary terms can help to avoid the hijacking problem. Section 4 explains the action likelihoods of the motion and structure of grouped object neighbors. Section 5 concludes the paper and discusses about future research directions.

2 Particle Filters for Tracking Multiple Targets

For the simulation of tracking multiple objects, we detect dynamically moving multiple touch points in our tabletop display system as shown in Fig. 1. The touched fingertips are obtained through an infrared camera capturing the scattered infrared rays from the acrylic display [8]. Using the segmented binary image of fingertips we detect the touch points using 8-neighbor labeling algorithm. The result of detection is the list of ID, location and size of each touch point but the ID of each touch point is not guaranteed to be same in upcoming image frames [5].

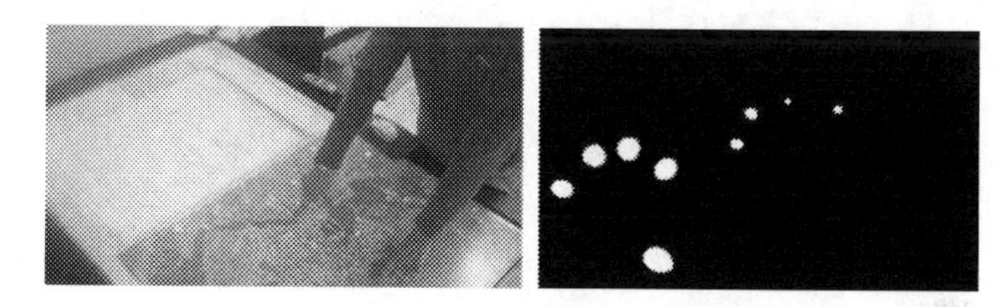

Fig. 1 Simulation system of object detection of touch points based on our tabletop display

Fig. 2 shows that the identities of detected objects change within two consecutive frames. Using a labeling method, it is not possible to assign proper identities to objects consistently in consecutive image frames. The data association of detected objects maintains the identities of multiple touch points throughout a video sequence. The easiest way of data association simply attaches a tracker to each target. In this paper, we prefer to use independent particle filters for tracking each object since particle filter is very robust in nonlinear and dynamic environment. For tracking each object identity, we follow the process of particle filtering.

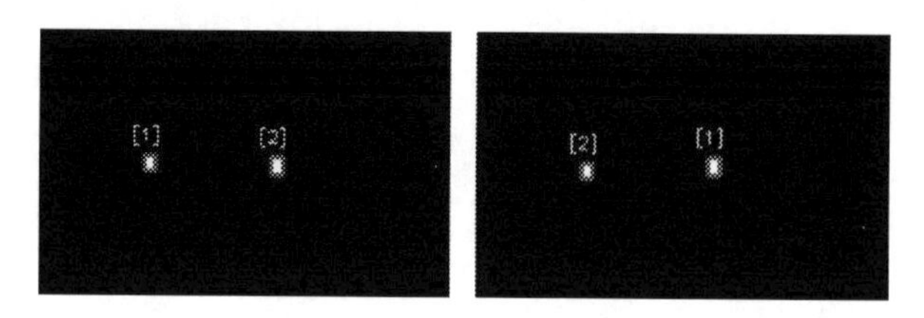

Fig. 2 The identities of detected objects using 8-neighbor labeling method can change

Particle filter is a implementation version of Bayesian filter [3], which estimates the posterior distribution by updating previous posterior distribution with discrete samples (particles). The benefit of Bayesian filter is the recyclability of previous result by adding current likelihood and prior into the previous posterior as like this

$$p(x_t|Y_t) = \alpha p(y_t|x_t) \int p(x_t|x_{t-1}) p(x_{t-1}|Y_{t-1})$$

where, x_t is the object state (location) and $Y_t = \{Y_1, \ldots, Y_t\}$ is the vector of observations which are the consecutive segmented binary images. The current posterior with all observation Y_t is updated from previous posterior $p(x_{t-1}|Y_{t-1})$ using prior $p(x_t|x_{t-1})$ and likelihood $p(y_t|x_t)$.

Particle filter has the prediction, evaluation and resampling steps. In prediction step every particle is predicted from previous particle set as like

$$p(x_{i,t}|Y_{i,t}) \approx \left[x_{i,t}^{(s)}, w_{i,t}^{(s)}\right]_{s=1}^{N}$$

where, $x_{i,t}^{(s)}$ is a particle, $w_{i,t}^{(s)}$ is the particle's weight and N is the number of particles.

The particles are proposed by below proposal distribution (transition model)

$$x_{i,t}^{(s)} \sim p(x_{i,t}|x_{i,t-1})$$

In evaluation step, every particle's weight is determined by its likelihood to update the predicted posterior to current time:

$$w_{i,t}^{(s)} = p(y_t|x_{i,t}^{(s)})$$

After updating the posterior from previous one, each target's position can be estimated by particle set of ith particle filter. The expected value $\overline{x_{i,t}}$ of the particle set in i^{th} particle filter statistically gives a reliable and smooth position of ith object.

$$\overline{x_{i,t}} = \sum_{s=1}^{N} w_{i,t}^{(s)} x_{it}^{(s)}$$

For estimation of likelihood, we use the integral image of segmentation image which is shown in Fig. 3. The likelihood is related to the number of white pixels covered by the particle window. Therefore the number of counting those pixels in integral image reduces to 4 times. Then the likelihood using integral image $\grave{y}_t$ is

$$w_{i,t}^{(s)} = \beta\left(\grave{y}_t(x_{i,t}^{(s)}:1) - \grave{y}_t(x_{i,t}^{(s)}:2) - \grave{y}_t(x_{i,t}^{(s)}:3) + \grave{y}_t(x_{i,t}^{(s)}:4)\right)$$

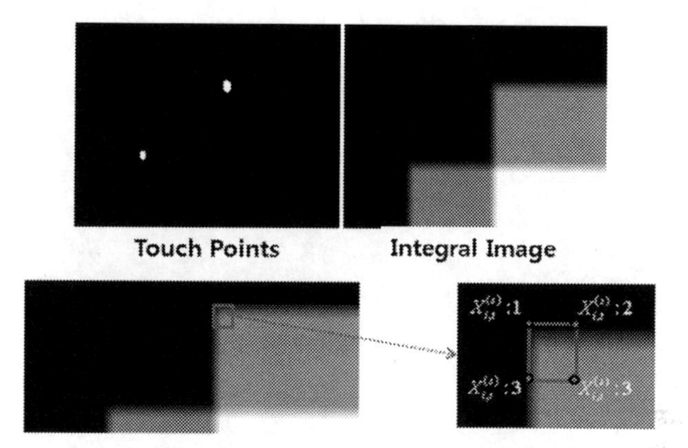

Fig. 3 The observation image (binary) is transformed to the integral image to calculate particle weights fast. The particle weight is the likelihood which is relative to the amount of white pixels within the location of particle which appearance can be seen here as a rectangle.

where, $\acute{y}_t(x_{i,t}^{(s)}:1)$, $\acute{y}_t(x_{i,t}^{(s)}:2)$, $\acute{y}_t(x_{i,t}^{(s)}:3)$, and $\acute{y}_t(x_{i,t}^{(s)}:4)$ are pixel values at the apexes of the particle rectangle. The summation of values in a rectangular region is done by accessing only four pixels. Comparing the simple summation of all pixels in rectangle, as known in computer vision area, integral image-based summation saves time exponentially.

Fig. 4 shows the resultant locations of tracked objects. Only using original particle filters with a proper likelihood function can track multiple targets robustly. The gray cloud is the distribution particles and green circle is the expected position of the particle distribution. This works fine until two objects are not near each other. Next section represents how to minimize the hijacking errors.

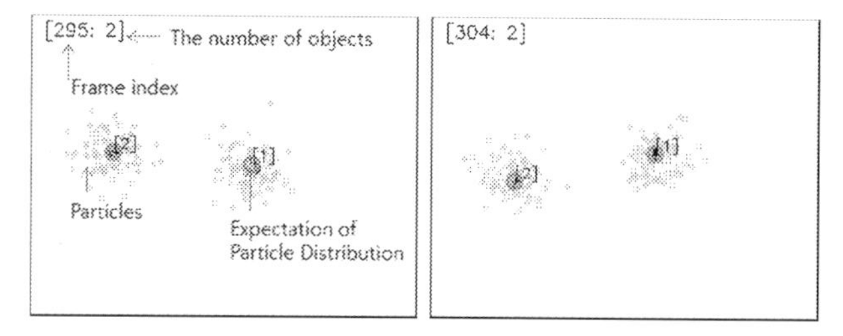

Fig. 4 Two frames (#295, #304) of tracking results in which the identities maintain consistently

3 MRF-Based Particle Filters

Independent particle filters are useful for single target tracking where no close and severe interaction happens between target objects. However when simultaneously tracking multiple objects, usually some of targets must get near or can cross over each other, then wrong target can hijack other tracking filters from neighboring object as shown in Fig. 5.

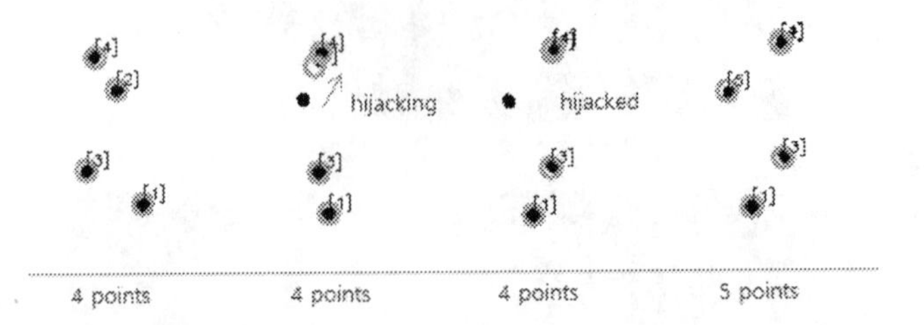

Fig. 5 As some objects get close their particle, Their weights can be wrongly augmented by other objects since the likelihood function has no guidance to the object identities

Hijacking other tracking filters on neighbors happens when some particles of the target are wrongly weighted by a near target. To avoid hijacking problem, Khan introduced a penalty function in the likelihood measurement using pairwise term of MRF motion model. The penalty function needs a graph for multiple objects. As a graph consists of vertexes and edges, in this multiple object tracking the graph has all gathered object positions as vertexes and connects them as links when the distance of objects is within the minimum edge distance as shown in Fig. 6. There can be several local graphs for the groups of locally neighbored objects.

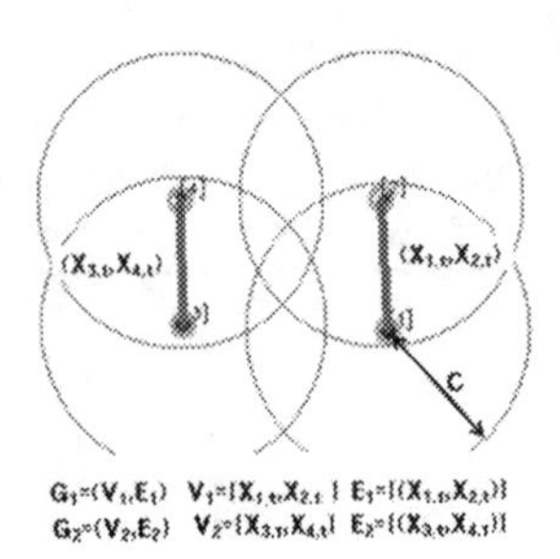

Fig. 6 Two graphs have been built based on the minimum edge distance *C*. Each graph has two vertexes and one edge.

The penalty function lowers the weights for those particles $x_{i,t}^{(s)}$ which are very near neighboring target $x_{j,t}$. If particle is getting near the neighbor, penalty value for the particle's weight is getting higher to reduce the effect of this particle in the tracking filter. The penalty function of particle $x_{i,t}^{(s)}$ with neighbor $x_{j,t}$ is

$$\varphi(x_{i,t}^{(s)}, x_{j,t}) = \begin{cases} \dfrac{C - \sqrt{\left(x_{i,t}^{(s)} - x_{j,t}\right)^2}}{C} & if \sqrt{\left(x_{i,t}^{(s)} - x_{j,t}\right)^2} \leq C \\ 0 & otherwise \end{cases}$$

where the minimum amount of penalty function is zero when the indexed particle is almost beside of a neighboring target. If the distance between the particle and neighbors is not less than C, penalty value is always one.

To apply the penalty effect, the particle is weighted with the likelihood function and the penalty function using

$$w_{i,t}^{(s)} = p(y_t|x_{i,t}^{(s)})\{1-\varphi(x_{i,t}^{(s)},x_{j,t})\}$$

Where x_{jt} is the neighbor which is in the minimum distance with the indexed particle. Fig. 7 shows the penalty effects on the particles and how different with Fig.5 to avoid the hijacking problem.

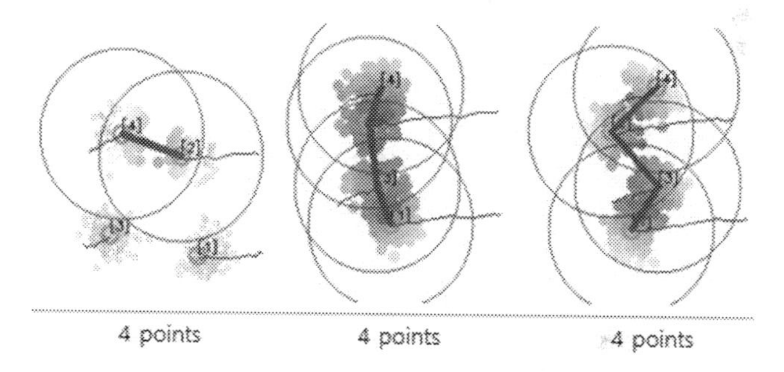

Fig. 7 The penalty effects are relatively colored as red on the locations of particles and by comparing Fig. 5 this approach can avoid the hijacking problem

However the time of the graph made is behind of one frame time. To eliminate the effect of one-frame late problem, we define a unary term to utilize the observation image. When the particles are positioned above neighbors, the likelihood function gives the maximum value. Therefore the previous penalty function needs to be changed to avoid the wrong maximum value of likelihood function. The unary term we define is used to check whether the particle rectangular area has white pixel or not and if it is near neighboring targets. The modified penalty function is

$$\varphi^*(x_{i,t}^{(s)},x_{j,t}) = \varphi(x_{i,t}^{(s)},x_{j,t})h(x_{i,t}^{(s)},x_{j,t}),$$

$$h(x_{i,t}^{(s)},x_{j,t}) = \begin{cases}\{1-u\left(\left(\dot{y}_t(x_{i,t}^{(s)}:1)-\dot{y}_t(x_{i,t}^{(s)}:2)-\dot{y}_t(x_{i,t}^{(s)}:3)+\dot{y}_t(x_{i,t}^{(s)}:4)\right)\right)\} & \text{if case } A\\ 1 & otherwise\end{cases}$$

where h is the modifier for penalty function. The case A means that the indexed particle is near neighbor and not near its target. In case A, if particle overlaps the neighbor not its target, the value of the modifier h is zero. u is a step function. Fig. 8 shows the effects of the modifier.as blue color.

Based on the penalty function with unary and pairwise terms, it is possible to reduce the hijacking occurrences. The modifier function helped the proposed method effectively avoid hijacking problem.

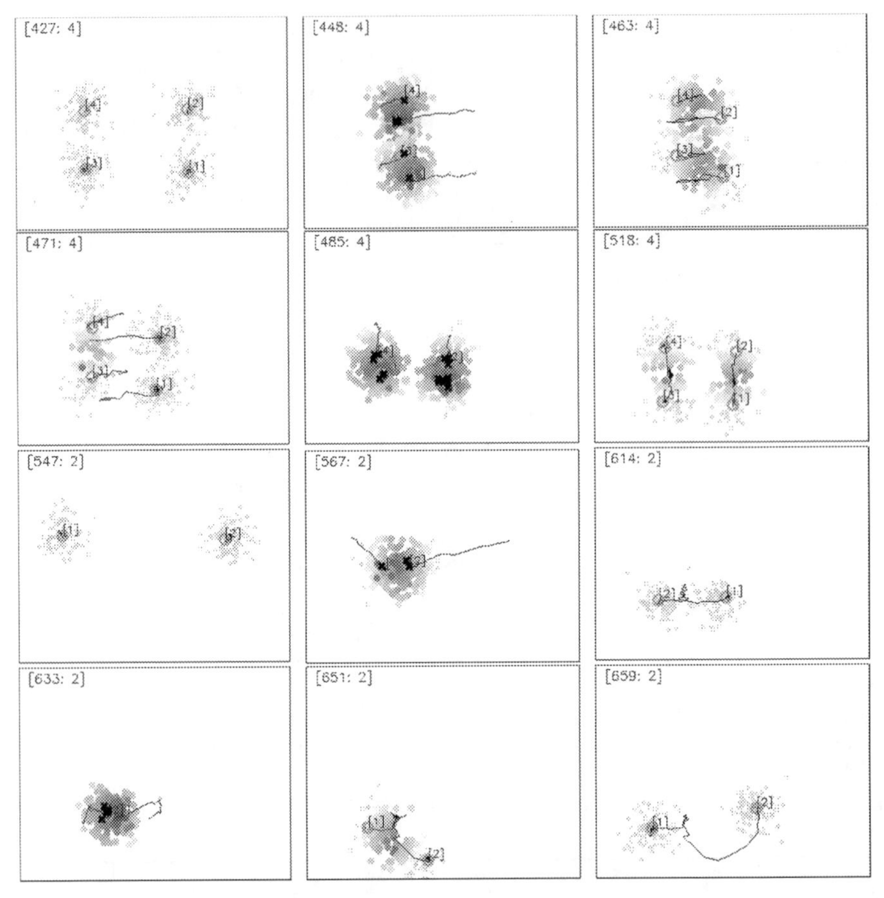

Fig. 8 The effect of the unary term-based modifier function shows that the particle with wrong maximum likelihood since the particle is located on a neighbor. And those particles are marked as black × and will have zero weight.

4 Action Likelihood Estimation

The trajectory of each graph shows the motion information of grouped objects such as translation, rotation and scaling with the description of the chord information of touch points. The chord information means how many points are moving or stable considering the structure of graph vertexes. The potential functions of the graph have the action information of multiple touch points. Therefore using the potential functions *(f1, f2, f3)* as shown in fig. 9 it is possible to establish action likelihood as

$$p_{transition}\left(f_1 | G_{i,t-1:t}\right) = \prod_{(X_i,X_j)\in E}\left[\frac{(dx,dy)_i (dx,dy)_j^T}{|(dx,dy)_i||(dx,dy)_j|}\right],$$

$$p_{scaling}\left(f_2 | G_{i,t-1:t}\right) = \prod_{(X_i,X_j)\in E} u\left(\left|d_{i,j,t} - d_{i,j,t-1}\right| - S_{MIN}\right),$$

$$p_{rotation}(f_3|G_{i,t-1:t}) = \prod_{(X_i,X_j)\in E} u(|\emptyset_{i,t} - \emptyset_{i,t-1}| - R_{MIN})$$

where u is step function and the translation likelihood $p_{translation}$ is the normalized inner product between all displacement vector, the scaling likelihood $p_{scaling}$ means whether all points are scaled larger than S_{MIN}, and the rotation likelihood $p_{rotation}$ means whether all points are rotated larger than R_{MIN}.. These motion likelihoods can be measured concurrently.

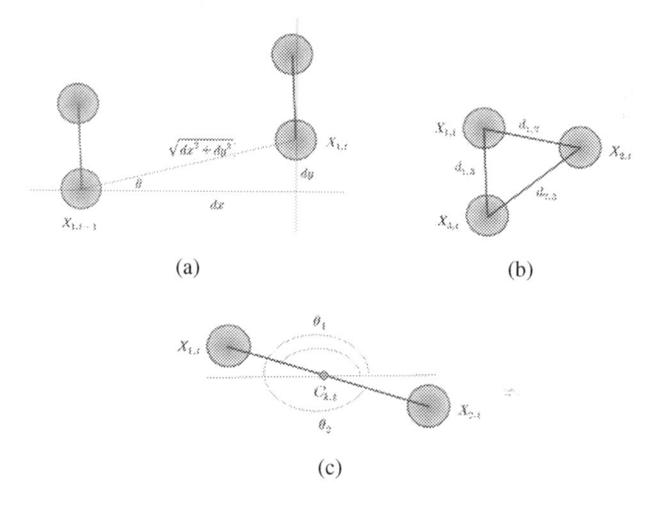

Fig. 9 the potential functions (f1, f2, f3) captures the motion information of vertex in two sequential time steps. (a) $f_1(X_{i,t}, X_{i,t-1}) = [dx, dy, \emptyset]^T$ (b) $f_2(X_{i,t}, X_{i,t-1}) = d_{i,j}$ (c) $f_3(X_{i,t}, C_{k,t}) = d_{i,j}$

Another potential function $f_4=[n, n_{move}, n_{stable}, n_{in}, n_{out}]^T$ represents the chord information of a graph. n is the number of touch points. n_{move} is the number of moving touch points, n_{stable} is the number of nonmoving points, n_{in} is the number of incoming points and n_{out} is the number of outgoing points between time t-1 and t. Considering the number of f_4 parameters five chord likelihoods can be defined by

$$p_{chord\_N}(f_4, N|G_{i,t-1:t}) = \int \delta(N - n),$$

$$p_{chord\_N_{move}}(f_4, N_{move}|G_{i,t-1:t}) = \int \delta(N_{move} - n_{move}),$$

$$p_{chord\_N_{stable}}(f_4, N_{stable}|G_{i,t-1:t}) = \int \delta(N_{stable} - n_{stable}),$$

$$p_{chord\_N_{in}}(f_4, N_{in}|G_{i,t-1:t}) = \int \delta(N_{in} - n_{in}),$$

$$p_{chord\_N_{out}}(f_4, N_{out}|G_{i,t-1:t}) = \int \delta(N_{out} - n_{out}),$$

where N, N_{move}, N_{stable}, N_{in} and N_{out} are user queries on the graph, and δ is delta function; $\delta(0)$ is ∞ or otherwise zero.

We have tested all action likelihoods and the estimated likelihoods and chord information in several motions as shown in Fig. 10. With all defined action likelihoods (motion, chord), any multi-touch gesture command in real-time application can be defined as a joint likelihood with them. For example, when we simply

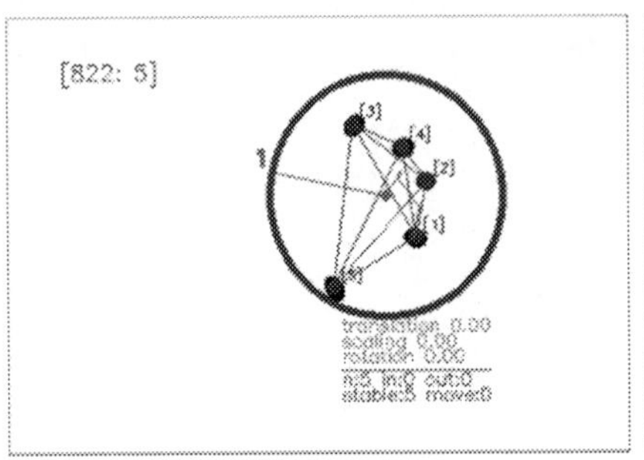

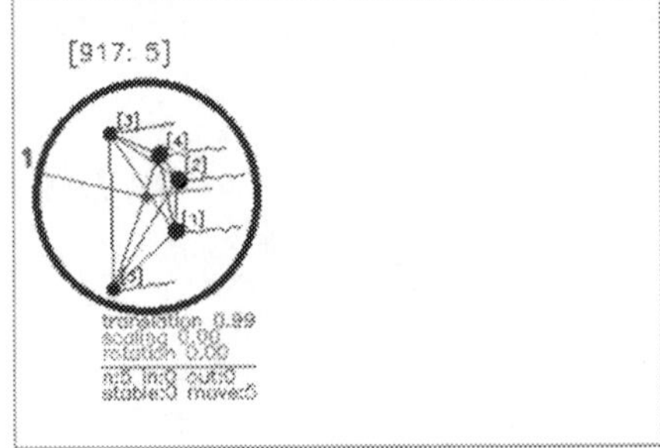

(a) translation

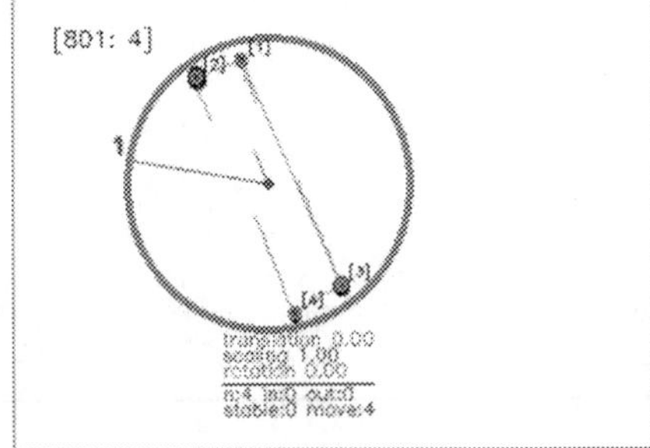

(b) scaling

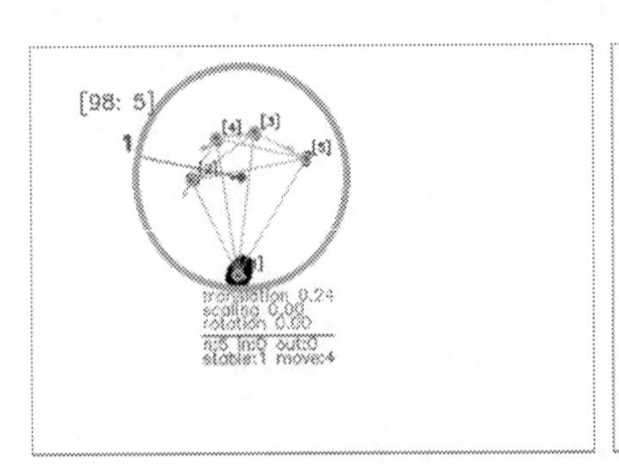

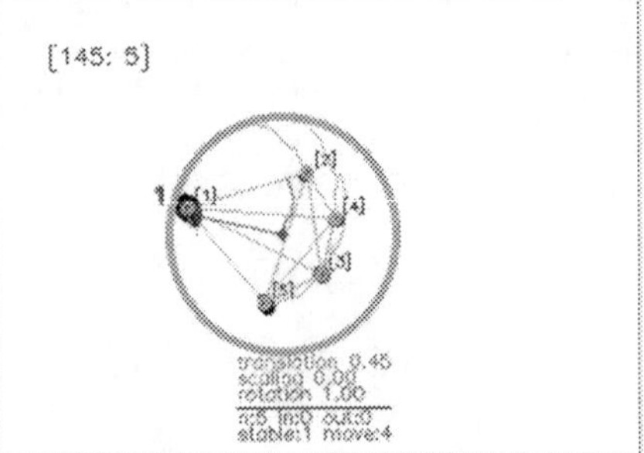

(c) rotation

Fig. 10 The action likelihoods of translation, scaling and rotation, and the chord information of $f_4=[n, n_{move}, n_{stable}, n_{in}, n_{out}]^T$ are shown below graph

use a multi-touch action (drag with two touch points) for mouse-drag command, the mouse-drag command can be defined as

$$p_{drag\_command} = p_{motion} p_{chord\_N}\left(f_4, 2 | G_{i,t-1:t}\right),$$

$$\text{where} \quad p_{motion} = p_{translation} p_{scaling} p_{rotation}$$

5 Conclusion

When we use independent particle filters for multiple object tracking, some target objects can lose their trackers due to the influence of the wrong maximum likelihood of neighboring target objects. This is called the hijacking problem and usually happens when target objects gathered into a small area. To avoid the hijacking problem the proposed method is MRF-Particle filters where MRF motion model is combined with particle filters to lower the weights of particles. If particles are very close to the neighbors then those weights are reduced by the penalty function of both pairwise and unary terms to avoid hijacking problem.

Due to the weakness of the pairwise term defined from the outdated graph, we additionally define a unary term to compensate the penalty function using given the observation under present time. The unary term captures the wrong maximum likelihood obtained from the neighbors. By reducing the effect of the wrong maximum likelihood, we can reduce the hijacking occurrences. This evaluation will be conducted in future works.

Additionally we expect that the information of MRF graphs can be used for action recognition. We have proposed how to define action likelihoods and joint likelihoods with potential functions which define some action elements and chord information. Some basic information of actions can be measured as shown in the resultant images. Future work will include extensive gesture recognition works based on these user-defined action likelihoods.

Acknowledgement. This research was supported by the MKE(The Ministry of Knowledge Economy), Korea, under the ITRC(Information Technology Research Center) support program supervised by the NIPA(National IT Industry Promotion Agency)" (NIPA-2013-H0301-12-3005).

References

1. Reid, D.: An Algorithm for Tracking Multiple Targets. IEEE Trans. on Automation and Control AC 24, 84–90 (1979)
2. Bar-Shalom, Y., Fortmann, T., Scheffe, M.: Joint Probabilistic Data Association for Multiple Targets in Clutter. In: Proc. Conf. on Information Sciences and Systems (1980)
3. Schulz, D., Burgard, W., Fox, D., Cremers, A.B.: Tracking Multiple Moving Targets with a Mobile Robot using Particle Filters and Statistical Data Association. In: IEEE Int. Conf. on Robotics and Automation (2001)

4. Khan, Z., Balch, T., Dellaert, F.: MCMC-based Particle Filtering for Tracking a Variable Number of Interacting Targets. PAMI 27(11), 1805–1819 (2005)
5. Viola, P., Jones, M.: Robust Real-time Object Detection. Int. Journal of Computer Vision 57(2), 137–154 (2002)
6. Islam, M.Z., Oh, C.M., Lee, J.S., Lee, C.W.: Multi-Part Histogram based Visual Tracking with Maximum of Posteriori. In: Int. Conf. on Computer Engineering and Technology (2010)
7. Stalder, S., Grabner, H., Van Gool, L.: Cascaded confidence filtering for improved tracking-by-detection. In: Daniilidis, K., Maragos, P., Paragios, N. (eds.) ECCV 2010, Part I. LNCS, vol. 6311, pp. 369–382. Springer, Heidelberg (2010)
8. Han, J.Y.: Low-cost multi-touch sensing through frustrated total internal reflection. In: Proc. 18th Annu. Assoc. Comput. Machinery Symp. User Interface Software Technology, pp. 115–118 (2005)
9. Oh, C.M., Lee, Y.C., Lee, C.W.: MRF-based Particle Filters for Multi-touch Tracking and Gesture Likelihoods. In: 11th IEEE International Conference on Computer and Information Technology, pp. 144–149 (2011)

Author Index

Printed in the United States
By Bookmasters